THE WAR IN MAN
Media and Machines

THE WAR

Media ar

IN MAN

Machines

by Frederick D. Wilhelmsen and Jane Bret

University of Georgia Press, Athens

MACHINES

TIME

THE SYMBOL

WORK AND
LEISURE

CONTENTS

The manuscript of this book was encoded using an Optical Character Recognition font on a Model 71 IBM Selectric typewriter by the University of Georgia Press. The OCR typescript was scanned by a Control Data Corp. electronic scanner and put on computer by Aspen Systems Corp., Pittsburgh. Editing and proofreading was done by the Press from a computer print-out furnished by Aspen Systems. After corrections were made on the magnetic computer tape, type was set by Aspen Computype, Inc., St. Paul-Minneapolis, on a Harris-Intertype Fototronic CRT. Finished, made-up pages were received from the CRT which was controlled by a computer program devised by the University of Georgia Press and Aspen Systems Corp. The book was printed and bound by Lithocrafters, Inc., Ann Arbor, Michigan.

Machines | The War Between the Two technologies

"**MY** ultimate criterion for a man is whether he is soft or hard on the Wheel." This quixotic gauntlet thrown in the face of the modern world is advanced by a friend of the authors of this study, a Southern squire who boasts that no one in his family as far back as Adam has ever been in "trade." His war against the wheel might well stand as a flag as round as a ship's helm, a circular symbol dividing all men within our world since the Renaissance. A famous Englishman swore that he would tramp to Rome on foot and that he would use no "wheeled thing," but he made the last lap down the long spine of Italy on a cart with his feet dragging on the road behind—thus fulfilling with feet the vow's letter and with his behind making obeisance to the wheel, hated with proper metaphysical vigor but respected from the very bottom of being.

Some men warm to horses and others like carts. And then some others prefer automobiles because their wheels go faster. There are hearts that expand to "the wheels of industry and commerce" and there are other hearts that expand not at all. The wheel is a splendid symbol. Wars and civil wars, factions and disorders, doctrinal conflicts that went to the roots of the spirit have set brother against brother and father against son in bloody combat from the Roundhead victory over the Cavaliers in seventeenth-century England—when the anti-wheels lost—to Vietnam in the twentieth—where the pro-wheels seem to be getting, for once, the worst of the deal.

The Southern Tory is hard on the wheel whereas Karl Marx was soft, but both are in agreement that the wheel—an apt symbol for technology —has made the world go round. Until very recently a man's ultimate commitments could be discerned by his reaction to technical progress, by the priority he gave it, and by the place it assumed within his hierarchy of values. Therefore we begin with the wheel, not in order to divide the men from the boys—that is not our intention—but with the hope that the wheel might teach us something about technological thinking, the nature of machinery, and its influence on man's history.

When God or Nature or Evolution or the Omega Point first deemed it proper that man make his appearance on this globe, there were no wheels around. The circular form was first discovered in things, possibly in trees or in sun and moon.

Nothing constrained our ancestors to think about the wheel. We have the massive testimony of the highly-developed Inca civilization in Peru. It never knew the wheel, and this despite a magnificent irrigation system that spanned hundreds of miles connecting a dust of villages that knit into unity one of the most formidable empires the world has known. Hitting upon the idea of the wheel by abstracting it from circular things in nature was not forced upon mankind as a necessity. The wheel was discovered by a wit who saw a connection between nature and man's needs. Abstraction made the wheel possible. Abstraction involves two moments: one which disengages some form— triangularity or circularity, for example—from the complex density in which it exists in nature, and one which suppresses mentally that very density. I cannot grasp wheel as wheel if I concentrate upon oak or pine. I cannot understand triangularity if I insist upon thinking about the chalk in which this form is drawn on a blackboard. I cannot think seriously about a rectangular prism if my attention is fixed upon gold in Fort Knox. Abstraction, therefore, both takes something into, and leaves something out of, the mind. Abstraction, which yielded the wheel in a primordial past, is typical of technological thinking as such. Abstract thought which gave birth to the rationalization of work demands an initial analysis of a whole into its constituent parts: the oak or the pine tree is subjected to an intellectual penetration that eventually reveals the trunk as round. Reality in its very existence in nature is not shattered into its parts: rectangular prisms are one with gold; triangles on blackboards are one with chalk; circles are one with wood. Things that exist in nature are unities, and this implies the synthesizing of their "parts" into an existential density grasped by the mind and sensibility only when it contemplates the whole. The primordial condition of technological thinking is a principled or systematic forgetfulness of the whole in favor of the part. This principled

2

forgetfulness by the mind is best described as an act of violence because the contrary of violence, respect and reverence, is always conferred upon a whole, be that whole personal or not. Whereas it would be Manichean to agree with Berdyaev that all abstraction is sin (violence), it is nonetheless true that every violent act involves abstraction, the very condition for the thinking which produces machines. For this reason, possibly, rapists and near-rapists look upon women as sex machines.

Now violence is not necessarily bad. Otherwise, we would have to forego swatting flies. Nonetheless the evil of moral violence is already implicated in the cerebral violence involved in every act of abstract analysis. Abstraction—Marshall McLuhan calls it "fragmentation"— is the dispersing of wholes into parts, shattering unities into multiplicities. Violence in the streets today as reflected through the mass media has its roots in a mind which is thoroughly conditioned by technological thinking. Gabriel Marcel sees this type of abstract thinking as one which turns both personal and impersonal subjects of being, of what we have called synthetic unity, into objects pinned to the screen of consciousness. Butterflies dissected by the mind form the very condition for analytic fragmentation of the real. The bulldozer that pulls up trees along with weeds, that levels slopes and hills, that destroys nature's unity, is an apt symbol for the analytical and fragmented mind projected into reality. The bulldozer could not synthesize any more than an elephant could put back into place with its toes the pieces that it had swept off a chessboard. It matters not at all that the bulldozer is an instrument in the service of man. Our judgments here are not moral but phenomenological. The machine violates nature as it carves it into pieces which are later reassembled —on the assembly line. This *structural* violence done the real is one with classical modern technology. It is a form of rape, even though the rape in question often be innocent and in the service of man.

The power of machine technology to transform physical nature was always known to the West, but it was only released fully during the Renaissance. The atomization that followed caused the destruction of older vertical hierarchies within society, the transfer of economic power from smaller to larger units, the drying up of old culture, the stripping away of economic independence. As mathematicized science subsumed more and more phenomena under increasingly simpler theories, hoping thus to explain the universe through a handful of axioms, so too did the technology created by this science aim at an ever-increasing simplicity of function: multiple units of production

3

were replaced, whenever possible, by single units; multiplicity of trades and occupations decreased as time moved on; embattled workers united in unions that resisted the supposedly predatory cupidity of capitalist greed, but they dissolved with hardly a whimper when technology eventually got around to absorbing them within itself.

The peasant and the homesteader were rendered superfluous in nation after nation. The special skill of the craftsman withered. The small businessman closed up shop because he could no longer compete. Hundreds of occupations, traditional ways of life defining dozens of classes for centuries, simply ceased to exist. Millions of people, finding their old places in society destroyed, their very history annihilated, were absorbed into the masses where their loneliness and resentment made them easy prey for totalitarian mass movements.

Sir Arthur Eddington stated the goal of modern science when he wrote that "science aims at constructing a world which shall be symbolic of the world of commonplace experience." This symbolic world is constructed so that the "commonplace" world can be manipulated. But in order to manipulate the real, science needed instruments—machines. And in order to get them science had to absorb a good deal of that very world within the new world of industry—technics—where things no longer were what they had been, but were rather technical means to technical ends. Thus a philosopher of science, commenting on the world of Dr. Einstein, concluded that behind scientific theory and technical organization there rested nothing "but a skeleton structure of symbols," of abstractions doing duty for realities.

Nature thus entered the field of the new mathematicized science in the Renaissance only so far as it could be stripped of its non-mathematical characteristics, or, more accurately, only so far as these could be symbolized mathematically. By suppressing the organic and the qualitative, the world could be systematized, rationalized, transformed, mastered. The secret of a total power over the real became man's refusal to look things in the face and see them as they are. When man averts his total gaze from things, he can reach out and possess them for himself, just as does the purse snatcher on a crowded street: if he saw the woman in the totality of her life and of her history, he would think twice before snatching that fragmented item known as a purse. This again simply illustrates the essence of violence.

Paradox though it might be, the truth is brutal: scientific power over a thing is possessed in proportion to our failure, often a principled failure, to know that thing as it exists. Conversely, a thing is known to be as it is when this power is relinquished and when the

thing is contemplated in its very existence. The dream of total technological control over reality has progressively removed mankind from
things as they are. It emptied the soul of its hidden reserves of loyalty
to the being of the world even as it turned the soul to a new order of
things. This dream threatened the social order even while it promised
a paradise on earth.

The older order, sealed in Greek philosophy and Christian ethics,
did not take kindly to the advent of machine technology. Content so
long as technics restricted itself to the modest disciplines of astron-
omy, optics, and music—sciences that comforted navigators at sea,
the shortsighted at home, and the devotees of Apollo at the harp—the
older Europe was not prepared to permit the new science to transform
radically the human community. Possibly a dim symbol of this suspicion of the machine is found in a tale told about Albert the Great,
himself a man learned in the curious and natural sciences. Confronted
with a life-sized doll that could walk, a precursor of the robot, Albert   5
smashed it with his staff. This spontaneous reaction—be it history or
myth—bespeaks man's ancient fear that machines will one day render
him obsolete. The classical and medieval worlds located the morality
of technics within the more familiar tensions existing between art and
morality. Technology, according to the tradition dominating the West
until early modern times, was essentially nothing more than the imposition of mathematicized form on matter. It followed that technics be-
longed as thoroughly to the artistic order as did painting or sculpture.
Both re-worked the raw material of nature. This tradition inherited
from Aristotle the conviction that the end of all art is the goodness of
the thing made and not that of the maker. (Were it otherwise—argued
the Aristotelians—a man could not ruin his health in the service of his
art: the existence of Bohemia and of Greenwich Village would seem
to support the Aristotelians on this point at least.) If the good of art is
to be subordinated to the good of man—and the ethics in question
dictated that it be so subordinated—this can be achieved only by the
Aristotelian virtue known as prudence, by man acting politically and
socially in a responsible fashion. Art cannot, so went the reasoning, be
controlled by prudence from *within art itself.* Whenever a man has
tried to do so he has irrevocably ruined his art. The painter cannot
paint with his good will. The poet cannot write with his moral probity.
The engineer cannot build with his decency. The nun cannot teach
with her piety. Conversely, the common good cannot be determined
by an artistic judgment. Technology, insisted the tradition, is incapable of saying anything about the goodness or badness of the social or-

der. The conclusion was inevitable: technology—a neutral instrument—ought to wait humbly on the intelligence and will of its creator who can use it or not; machines (according to the classical-medieval tradition) ought to be like pocket watches which can be taken out of pockets and put back in them at the pleasure of their owners.

The reasoning of the older teaching had inevitable social and political consequences. Society had the right and duty to lay down the law to science in all things affecting society's basic cultural structure. Should technological advance threaten to wipe out a community enshrining a way of life considered by the tradition to be inherently good, then that community had the right to step in—with the sword if necessary—to protect its existence. No piety to the scientist because all piety was already lavished upon artisan and peasant.

The scholastic position on technology has been revived in our time by philosophers such as Jacques Maritain who tend, however, to see the clash between art and morality as revolving around the fine arts and their possible repercussion on sexual behavior. The contemporary clash between art and morality has centered around a supposed right by the community to *censor* books, movies, statuary, and paintings whose significance damages or is thought to damage the moral tone of the citizenry. The far vaster problem of society's moral control of the technological problem has hardly been discussed at all. But the historical drama has had little to do with peripheral issues such as the incipient pornography found in French impressionism. History has placed the battle where it squarely belongs: between morality and technology. The Aristotelian and medieval tradition was not content initially to repeat its convictions simply in textbooks. Although Machiavelli insinuated sweet promises of total power into the minds of Renaissance princes, there were others for whom the inherited teaching was the stuff and substance of morality itself. The new order and the old tested their strength in England. The old set up the royal standard in Nottingham and called into its service the loyalty of peasants, yeomen, and nobles who remembered an older epoch; the new rallied around Parliament, shop keepers, and ship owners, and was charged with the nervous energy of the southern counties. The older convictions about the role of technology in life had been incorporated in the English statute running from late medieval times, through the reigns of Edward VI and Elizabeth I down to that of King James II, last ruling monarch of England. A number of examples will suffice to demonstrate the thesis. The fulling mill in 1542, the gig mill in 1551, and the tucking mill in 1555 were outlawed because they were regarded as working toward the "final undoing of the industry concerned." The quaintness of the examples makes us smile, and they almost suggest Captain Ludd. But the antiquarian flavor ought not to cloud our understanding of the essential issue. The English kings were convinced that each man had a right to pursue the trade he had inherited, that each man had a right to the living he had made his own, that all orders and classes forming the hierarchy of English society were as natural to existence as was the coming up and the going down of the sun or the rise and fall of the tides. The English kings—men whose consciences had been fashioned totally by the old order—were the Walter Reuthers of their time, complaining bitterly not about automation in the automobile industry but about incipient automation in, for example, the garment trade. That most contemporary labor leaders belong to the conventional "Left" whereas the Stuart kings are considered archetypes of the conventional "Right" is, of course, utterly irrelevant to the issue, although it does illustrate the fallacy of the misplaced consequent.

Suspicion of technological advance was not then the result of any romantic longing after "the good old days" any more than romanticism is today a principle guiding the decisions of the teamsters union. As both attitude and pose, romanticism in the sense of a celebration of the old belongs to the nineteenth century and is practically unknown today. The critical issue is the concrete attempt of governments then and labor leaders today to protect the interests of the working communities they represent. Given that governments today tend to represent science and progress rather than working men, the role of the kings has passed to the leaders of what older generations called guilds and of what we call unions. Today the working community is not interested in progress. Its goal is security in the things that it has and knows. When the poor were in the saddle under the reign of Charles I, the limitation of technics was simply part of a broad, social legislation that forbade, among other things, the purveying of counterfeit jewelry (April 18, 1636) and the fraudulent sale and packing of butter (Nov. 13, 1633), that warned starchmakers and malsters that their product was not as necessary to human life as was the raw material of their industry; and that decreed, concerning workers in cloth and yard, "the increasing of the poores wages labouring therein." As far as technological progress was concerned, the king in Whitehall and the poor of his realm considered it to be akin to fraud.

But the new machine technology—resisted by crown and cottage— was infused by the new Puritan ethic which invested pure work with a mystique completely unknown in previous times. Given that pure work was better served by the new technology, the Puritans—the new middle class—threw their weight into the service of mechanical progress. This attitude contrasts sharply with the reaction typical of artisans then and of blue collar workers today: for both, industrial progress bought at the price of technological unemployment is not a virtue and is no sign at all of the Coming of the Lord: it is rather a well-honed definition of sin. This attitude, common today in a working population that fears automation and in a "drop-out" generation of youngsters that simply "hates those machines," as McLuhan puts it, was once that of the Stuart kings against the new capitalist class that seized upon technological advances, financed them, and reaped the profits accrued from them.

We have come a long way from the times of Charles I of England and Scotland. Since the death of the somber and lonely Stuart king on a balcony at Whitehall, sent to the block by the new rich, the "saints," and men like Praise God Barebones, the West has lived through the Calvinist revolution which made possible the Industrial Revolution in the late eighteenth and nineteenth centuries. The whole movement was transfigured by the Puritan passion to conquer the world and thus justify man to God. Herbert Marcuse in *Eros and Civilization* calls it the victory of the production principle over the pleasure principle. In any event, the older classical and Aristotelian politics, which would have controlled technics, disciplined industry, relaxed humanity in the name of contemplation and play, and subordinated scientific progress to immediate human goals and aspirations, rapidly gave way to man's romance with the machine. As Gabriel Marcel has written, it is easy to take up technics; it is almost impossible to lay them down. Once a man has assumed what is both a burden and a promise, he will never be the same again. The Pied Piper of Power, the high priest of machine technology, might very well trip over the edge of a cliff and thus destroy himself; but history has demonstrated that wherever he leads, mankind will follow, and this despite the fears which whistle through labor halls today and which cascaded down the halls of the royal courts of yesteryear.

We are called upon, therefore, to penetrate the nature of the machine in its impact upon the human mind and sensibility. Technological thinking as well as its products initially attempts to disguise its remoteness from things human and organic. Or, more accurately, men disguise this remoteness because they secretly fear a technological future until it has become a past, when it then converts itself into a quaintness which can be handled either snobbishly or comically by sensibility and will. The first automobiles looked like carriages, and it was not until 1946 with the Mark V that the Jaguar dropped its detached headlights which were still referred to as "lamps." As new machines gradually enter into the fulness of their perfection they shed the superfluous appendages of older and obsolescent models. They shake off the vestiges of the organic, the non-rationalized, the non-abstract. But man always seems to be lagging behind the machine in this regard, for he is reluctant to admit that the new is not simply a variant of older modes of craftsmanship. Fisher Body Company even staged a contest for boys and gave prizes for constructing model carriages, not cars. Rolls Royce has clung to its elegant radiator because it is an excellent status symbol, though it no longer has any real function. The Rolls Royce of mid-thirties vintage strikes an image more antique than that of the last U.S. square-rigged sailing ship, *The Star of Finland.* Old machines look far more ancient to us than do even older non-rationalized artifacts because the former bear the marks of the superfluous whereas the latter do not. A gold watch of the last century, its case all encrusted with rococco nymphs—fat, saucy, and sexy to an older taste—wound about with a lordly key, and giving off chimes upon the hour, is more venerable to us than are latter-day models of hour glasses and sun dials which were in use before the Pyramids were a reality. Hollywood captures this truth in its rendition of Jules Verne's *Twenty Thousand Leagues Under the Sea.* The twentieth century is at odds with the warm decor of an overstuffed Victorian salon, which Verne simply transferred from the London of the hansom cab and the meerschaum pipe of Sherlock Holmes to the interior of the first submarine. Audiences seeing the motion picture did not truly experience the future at all: they sensed rather that they were thrown backwards into the England of the last century. The superb cut-glass brandy snifters of the captain's bar, his brocaded dressing gown, the magnificent cigars housed in costly humidors and then cut with dainty knives fashioned for the purpose, the leather-bound books—none of this has anything to do with the spare and lean interior of today's submarine where asceticism weds the abstract perfection of technology. Even the submarine of Jules Verne's fancy struck its prey as does a swordfish, thus mingling images of an obsolescent form of fish life with dreams of a technologized future.

Man is truly what Parmenides would have called a "double-head" in his attitude towards machinery. He wants technology desperately, but when he gets it he covers it up and pretends that it is something

9

else. He keeps the lid on as long as he can, which usually means a generation or two. His vision, ocular as well as intellectual, is offended by admitting the new to the screen of his consciousness. Long after the steamship was perfected it still carried masts and sails. Long after the steamship was obsolete, motor vessels carried ventilators disguised to look like steam stacks. Naval officers and West Point cadets—engineers and technicians and radar specialists—leave church on their wedding day under an umbrella of swords. And we still retain the term "cavalry" for mechanized units of an army that last heard the hoofs of galloping horses when Black Jack Pershing launched his Mexican adventure before World War I. We readily accept mechanical progress as an act done, but we resist it as theater to be seen. The Old Guard is the human eye. The eye is the perpetual ally of the conservative spirit. In the past this spirit insisted on decorating maps with sea monsters and mermaids, all soft and lovely, thus offending the cold precision of the mathematical thought which went into the maps themselves. Technology struggles into its own by means of an internal dynamism which has little to do with human hearts and even less with human eyes.

Man's ambivalent attitude towards the machine is a consequence of an ambivalence in machine technology. Friedrich Georg Juenger's insistence that the pinnacle of technological perfection coincides with its failure, points to a built-in contradiction within machinery. The machine, initially an extension of the human body, progressively declares its independence of the body as it is perfected. The issue demands explanation.

Rudimentary machines extend man's legs and arms as do bicycle pedals and lawn mower handles. They elongate and strengthen all four human appendages. Had man come into being as an overblown centipede, his machinery would probably have been complex but personally manipulated. The division of labor, the critical principle of industrialism, would have been cancelled by an internal synthesis. Industrialism was the result of man's having only two legs and arms instead of some fifty each; it was the consequence of nature's being parsimonious with mankind.

Labor remained reasonably undivided and whole in non-rationalized communities where farmers and artisans moved from one simple machine to another. Each of them could be worked with hands and arms or legs and feet or both, but they could be worked only one after the other by the same man. When western civilization moved out of an essentially agrarian community within which the family formed a center capable of managing all the machines necessary to secure the needs of life, technology forced a further division of labor which went beyond that primitive division demanded by all space and time. Machines, initially extensions of man's modestly enumerated appendages, proliferated to the point where they could no longer be handled by individual men or even family units. Most machines were thus divorced from most men, even those engaged in machine production. The limitations of the human body—only four limbs—made possible the assembly line as conceived by that archangel of technology, Henry Ford I. Nonetheless, even within the dispersed and shattered structure of the factory, at first one man ran one machine and very often it danced to the tune he played. Differences between older sports cars of the same model are differences in the craftsmanship and hence in the personality that went into making them. Whereas these differences are weighted on the side of the positive in luxury automobiles, they tend to be negative factors in proportion to the removal of hand craftsmanship from mass production. An instance of an otherwise good model automobile, mass produced, which turns out to be a "lemon" is traceable usually to faults in the human element involved in its production. The logic of technical advance progressively eliminated the personal factor. Purgation of the human is one with mechanical progress. Machines, as they achieved the fullness of their perfection, were no longer "run" by men in any accurate sense of the term. They were rather "tended" as were boilers on steam ships which split the black gang into oilers and tenders as well as shovelers of coal. Machines that ran themselves needed from men nothing but an occasional ministration, a shot of oil here and there, a minor adjustment of dials and switches, etc. Men thus became the nurses of machines, the Florence Nightingales of technics. Factory workers were vitally important to the health of their patients but peripheral to their performance. This movement from active dominance over machines to simply doctoring them, coinciding with the growth of late nineteenth- and twentieth-century industrialism, transformed millions of men from active and often creative craftsmen into bedside attendants. This occultly feminine stance, itself offensive to the male, was rendered tolerable by a violent reversal of roles which restored to man a precarious masculinity. Man married the machine, his own mechanical bride in the sense of the title of Marshall McLuhan's first book. The subsequent mechanizing of sexuality was simply a consequence forced upon a society by men who were otherwise robbed of their manhood as they spent their lives taking machines' temperatures, feeling their pulses, and applying un-

11

guents. Psychologically, the machine *had* to be man's bride in order that it not be his husband. Whole generations of males could thus forget that they were little more than nurses. But they forgot at their own peril. The massive rejection of a technological homosexuality, itself a violence done the psyche, is partly responsible for sexual violence in our society, whose barometer rises everywhere in proportion to that of mechanical progress.

The machine—by an internal logic utterly beyond any moral control insisted upon by the older philosophy—declares its independence of man, although, as shall be demonstrated, the declaration never truly becomes reality. The issue is illustrated quite simply by the lawnmower, that admirable symbol of American middle-class suburbia. At first we pushed it and the thing did our will depending on strength of arms and keenness of eye as we surveyed our postage stamp empires. Tending the lawnmower—oiling and sharpening it—was instrumental to pushing. Then a little motor was added and this enabled us to push with less energy. Finally, the machine was perfected and we mounted the business, thus relieving legs of labor and arms of effort as responsibility for the operation retreated to the eye. Very soon buttons that could be managed by blind men will handle totally-automated lawnmowers. One of these charming devices was displayed on the Merv Griffin television show, where it appeared merrily running back and forth across the stage like a mechanized, souped-up turtle. Merv paid no attention to the thing once it was put into motion. In isolation from similar developments, the progress of the lawnmower looks like a liberation of man from toil. When the process is multiplied, however, it simply reveals all the more man's gradual transformation into the nurse of a world of machinery that works well on its own and that needs him only when it works badly. Man's superfluousness could not be more obvious, nor could the machine's anonymity.

Machine technology's *total* perfection, a mechanical universe that ran smoothly on its own without human intervention, would be self-destructive because machines in their primordial essence are simply extensions of man himself. It matters not at all that machines exist for men and that total mechanical perfection supposedly would be in man's service. Obviously, it would not! An automobile that ran itself without any interference from man would *not* be a servant. It would be a servant if it were controlled remotely by an electronic device, but in this case, a new element is introduced into technology which is not mechanical but electronic. The advent of electronic technology forces us to confront an ontological situation which differs radically from

12

that of the older mechanical order. So long as the electronic is an instrument of the mechanical it remains a subaltern, but the new age aborning is witnessing a reversal of this situation. The mechanical is losing its independence and is becoming more and more an instrument in the service of the electronic. The new technology forces us to look even more closely at the nature of the old, for as the old passes progressively out of history, civilization will find itself liberated proportionately from its subservience to pure process.

Machine technology is process attempting vainly to convert itself into independent existence. This is its nature and therein consists its inner contradiction. A machine is so constructed that it is a pure "being-for," a mechanism so formed that once turned from the use for which it was built, it is distorted. Try to imagine an automobile used for anything other than driving, a machine gun which did something other than spit bullets, a piston in the service of horticulture. Shunted from goals which are their internal dynamic principles, machines are either violated—pistol-whipping violates not only the man whipped but also abuses the pistol—or they are antiqued and thus declared *hors d'combat* , like Mississippi side-wheelers converted into restaurants or San Francisco cable cars decorating gardens. Machines, therefore, are so utterly functional that the world fashioned around them has tended to be purely functional. Its very art and style of life, its music and literature, as well as the rhythms of its psyche, reflect and thus symbolize the mechanical basis upon which our world rests. Nonetheless, mechanical process tries to declare its liberty and thus blindly but inexorably reaches for anonymity with an ever-increasing velocity toward its perfection. If a machine, however, were to convert itself into an independent reality, into being rather than process, it would look like a Rube Goldberg contraption. It would go round and round and *do* nothing at all. It would simply *be* and this *being* would reverse its teleology, which is *not to be but to do* . The Industrial Revolution based upon mechanics has avoided collapsing into being— sterile nonsense for machines—thanks to the advent of the new electronic technology. Given that this new technology can be bent to the old—as is ignition to the combustion engine—machine technology, as Juenger predicted, has not achieved that absolute excellence which would have been its failure. It is amusing to note that Juenger's famous book, published in the original German as *The Perfection of Technology,* was changed in its English edition to *The Failure of Technology* . They both amount to the same thing.

But Juenger did not take account of the newer electronic technol-

13

ogy whose phenomenology is pioneered today by Marshall McLuhan. Electronics finds its perfection, not in any instrumental services it might render machines, but in media which move information. McLuhan's thesis that "the medium is the message" not only properly compounds sign and signified, but it also provides an insight that can be expressed by stating that a medium is "being" rather than process or "being-for." Automation does not convert the mechanical into an independent existence that it can never achieve. Automation invades the mechanical world and renders it obsolescent because the distance between mechanical process and automated being is absolute; the chasm cannot be crossed. Show a savage an automobile and he will not know what it is for, what to *do* with it. Put that same savage in a room lighted by electricity and he will experience light, just as does the most highly civilized of men. The savage will find himself in a new state of existence, the state of "being-illuminated." He need know nothing at all about the nature of electricity: he will experience light just as he will a voice if he is handed a telephone or moving, talking images upon a screen if he is led into a theater. Machinery, on the contrary, is one with "doing." The steam engine, the piston, the cart, the automobile —all of them can *be* only to the degree to which they are used by men trained in their use. But nobody needs to know how to use electronic technical media in order that they be. Once they are "switched on" they produce modes of being which are indifferent to the use men make of them. Do with them what he will, man is now involved in a new order of existence. Electronic media, therefore, heighten something intrinsic to all media. There is a famous story about a clown who runs into town announcing that the circus tents are on fire and who is laughed at by the townspeople: whatever else he might do or say, the clown is first and foremost a clown. So too are all media.

Now electronic technology is antithetically opposed to mechanics in almost every aspect. McLuhan has demonstrated this brilliantly in his *Understanding Media.* Here we select one of these aspects crucial for our purposes. Mechanical technology tends towards anonymity as it approaches perfection; machines progressively throw off human intervention in their functioning the better they become, needing less and less immediate human guidance by their operators; they reach for an independence they never quite achieve; and they, in the measure of their excellence, depend less and less upon the human person. Electronic media do something entirely different. They involve the person by invading his entire nervous system and by eliciting his attention, willy-nilly. You cannot *not* be involved in the light that is on. Personal involvement is of the very meaning itself of electronic media, whereas personal involvement is a sign of primitiveness in mechanical technology. The issue can be expressed as follows: electronic media are independent in their own mode of existence, and they render man dependent by involving him; but mechanical technology, while dependent upon man, progressively attempts to escape this dependence. The results, of course, are antithetical. The electronic involves man more and more as it approaches its perfection. The war between these two technologies—the older approaching fullness at the very moment of its immanent obsolescence even while it mingles with the newer—has produced an intolerable tension in the psyche of western industrialized man.

The mechanical order has had four centuries now to depersonalize the human being and to render the world surrounding him remote and distant until he has almost come to sense himself a stranger in his own house. The new media, simultaneously, throw him back into what McLuhan calls "tribalization," into a new order of corporate existence that obliterates space and time even as it combines into unity by analogy a world in which men can no longer hide from one another. It is as though the machine were suddenly found out because somebody turned the lights on. Atomized by mechanical culture into solitary units that had been centralized in the gigantic factory which the West had become, cut away from one another by an ever-increasing division of labor, constrained to live in a world that was perforce as functional as was its technical basis—men suddenly found their private and depersonalized lives invaded by a gigantic Peeping Tom that synthesized once again the hitherto fragmented, that united the previously isolated, that abolished pure function in the name of a new communitarian being. We oversimplify for the purpose of illustration. Actually, the new technology has not abolished the old—not yet. It co-exists with the old, buttressing it peripherally even while gradually replacing it centrally. But in the meantime, *Western man is forced to live two rhythms which oppose one another.*

As a mechanical man, he is alone, anonymous, violent in the mechanization of his sexuality and resentful in the loss of his personality. As "beamed on" by electronic media which englobe him along with all other men, he is forcibly united to the whole, tribalized, reintegrated brusquely into a humanity he would otherwise avoid if he could. Machine tenders in Detroit, beer-swiggers at the *Oktoberfest* in Munich, scuba divers off Ecuador, space men on the moon and min-

15

ers in the bowels of the earth, the Pope in Bogota and Russian tanks in Prague on the same day—all are rendered simultaneous to a generation fed on television. Privacy as a stance and as a way of life is threatened. There will be no more invasions under the cloak of darkness, and even the slaughter of the Biafrans in Nigeria now works its tragedy into the front rooms of millions in America. No longer can anybody get away with something without being found out. Privacy—itself a necessary by-product of mechanical fragmentation—is today discovered, revealed, commented on. Princess Irene of Holland taking a sunbath on the roof of an Italian hotel in a bikini, discovered and photographed by prying news photographers, did not lose her privacy. It was not taken away from her. It was simply found out and violated.

Privacy, in the light of the new electronic media, has ceased to be a univocal concept: it has become analogical. Where time and space are not annihilated men can "escape." "Distance" belongs to nature and is inherent in walls and coves and secret groves. Electronic media do not destroy this discontinuity in nature which renders privacy possible. They simply absorb nature in such a fashion that what is distant and private *there* is no longer distant and private *here* inside the media; telemobility, a new term, simply means "far is near." In a natural sense a man can escape Washington and lose himself in Acapulco. If somebody calls him on the telephone from Washington, natural distance has not been annihilated: it has simply been rendered irrelevant—outraged. Nature is not destroyed by the new technology. Things remain just as natural as they were. But nature is now absorbed within a new "skin," rendering it, as McLuhan states, *one* among a number of environments in which man lives. The old mechanical technology can function only by altering nature—hence the valid suspicion entertained by romantics that machine technology is hostile to nature. The new electronic media do not alter nature. They locate it within a new border of existence. The two worlds war within the psyche of contemporary man in the war nobody talks about; but it is decisive to the battle being waged in the human heart, torn asunder by two hostile forces.

Electronic technology exacts a revenge that the enemies of the machine could never have dreamed of visiting upon the mechanized world they have hated with such bleak passion. Traditionalists and romanticists and even conservatives have spent generations detesting the slick impersonality and atomization produced by machine technology. The enemies of the wheel have deplored the drying up of craftsmanship and trades, the disappearance of the family farm, the ugliness of the big city, the cheapness of industrialized centralization, and the progressive impoverishment of life which resulted therefrom. These men used the benefits that accrued to civilization from the machine even while they plotted sweet treason against it. But they never imagined that the instrument of their wrath would itself be technological. They looked back nostalgically to the Middle Ages, but they did not know the Middle Ages were not in the past but in the future. The Duke of Cumberland, known fondly to all Scots as "The Butcher," the armed guard of the forces of English mechanical and economic progress, thought that he had stamped out the clan system in the Highlands in 1746 with fire and sword. But by the year 2000 the whole globe will be little more than one gigantic clan, with every man linked to every other man by a technology which for all practical purposes abolishes space and time and restores the familial and personal on a world-wide scale. The claymores and the bagpipes are coming home again, even if under a radically new form. The vengeance they will take on the mechanical past is still hidden in mists which cover the farther shores of history.

17

War itself is an instance of the clash between the two technologies. Bernanos in his *Man Against the Robots* pointed to the abstract nature of modern war by noting its tendency to divorce killing from direct responsibility. When soldiers charge enemies they can see, they are personally involved in the trench they take and the machine-gun nest they overrun. They sense fear and pain in the men they bayonet to death. This is their common sacrament of a shedding of blood, binding into one both victor and vanquished in a tragedy shared by both. German and Russian sharpshooters in the battle of Stalingrad were secret allies whose comradeship consisted in their having to kill one another. The duel unites and makes companions out of men otherwise indifferent to each other. This is why survivors of the duel always shake hands. But a technologized killer—abstracted from his enemy and incapable even of fashioning him into being within the imagination— the pilot of a bomber, say, over Dresden or Hiroshima in 1945— simply could not experience the violence he unleashed in dropping bombs. He could kill fifty thousand people and return to home base and a sandwich without in any sense of being involved personally in the enormous carnage he had caused. He knew no comradeship in death. He assessed no personal responsibility for what he had done. He shed no tear over enemies, who for him were nothing other than objectified targets pinned in symbols to instrument boards, registered as ciphers within a coldly-rationalized consciousness. As mechanical technology advances, violence grows greater, and personal involvement dims until it finally flickers out into nothingness. And all this reflects the pattern which governs all machines and the mechanical men who manipulate them.

Not only has the technologizing of war made soldiers less soldierly, as the military historian John Wisner insists, but this increasing anonymity today tends to be undone antithetically by the abolition of the limits of space and time achieved by electronic technology. The bomber pilot may or may not see the destruction he causes, but millions of television watchers do. They experience what he does but cannot see. Finding its very meaning in the moving of information, electronics become forms of corporate education in death. Let us think only of the millions who watched the two Kennedys carried off to Arlington and buried. Let us think only of the millions who in their living rooms saw Lee Harvey Oswald murdered by Jack Ruby. The man who stays at home gets a better picture of the war in Vietnam than the soldiers who fight it. The supposedly "passive" television watcher is clearly not so involved as is the man who is getting shot or doing the shooting. It would be absurd to insist that today there is no distinction between doing and watching, but there are signs that this situation is changing. There are signs that the distinction between observer and participant is disappearing. This is perhaps the most important single event in our transitional age.

A vivid instance of this was the 1968 Democratic Convention in Chicago which was revealed to be a relic of the age of pre-electronic media. The convention as an American tradition is based upon the democratic assumption that the ultimate unit in society is not the family but the countable individual. This political atomization was a simple consequence of the Industrial Revolution's fragmentation of reality into mechanical units moveable by rationalist manipulation. Countable individuals delegate powers which are then sub-delegated and re-delegated to ever smaller groups of individuals through established chains of commands. This dispersing of political action did not produce decentralization but an ever-increasing centralization in which power tended to be concentrated into platform committees, nominating committees, and all the other committees endemic to the classical American convention. The system worked splendidly just so long as the technical basis of society was itself mechanical. But when the hippies and the yippies and the New Left wanted to get into Convention Hall in mass in order to participate directly in its proceedings, they could not. Now there were a dozen immediate political reasons why the New Left wanted in and why the old party liners and the police wanted them out. But the most obvious reason why they *could not* have gotten in seems to have missed everybody. *There wasn't enough room!* In a conventional democratic situation in which society is fragmented into individuals in *classic chains of command,* the very limits placed on human action by space and time demand that power be delegated. This delegation of power even fragments the men who possess it. Delegate chairmen and others scouting for delegate support as they moved back and forth across the hall were less aware of the total picture of what was going on than were the candidates watching the show on television from their hotel rooms, and even less aware than ordinarily well-informed Americans everywhere seeing the same show from their living rooms at home. The walkie-talkies on the floor were the missing link between dinosaur democracy and whatever the future might portend. Old-fashioned involvement in politics demands a division of labor and a splitting of functions absent in the total involvement permitted the spectator through electronic media. Media destroy all chains of command, just as the jangling of the telephone cuts through and annihilates every going priority within home or office. In abolishing chains of command media destroy the flow of information from upper to lower and lower to upper echelons of participation. Many delegates had to wait for next morning's newspaper to find out what they had been doing the night before—even though millions of Americans watching television already knew.

19

There is something uncanny about this. The people who are most *in* politics, such as delegate chairmen for a candidate, are the most remote and *out* of it since the advent of mass media. Whereas the people who are *out* of it are the most *in* —sitting in their living rooms they can see all the delegate scouters *at once* and grasp simultaneously information which flows *out there* in a space-time sequence. The space-time sequence "out there" is simply abrogated "inside" the Tube. Houston and Detroit and London got a better view of the Battle of Chicago than did the police and hippies who fought it. And even though, in the case in question, the media were criticized by Mayor Daley and others for giving a one-sided picture of the street war, these criticisms were made known to the nation through media themselves, through Mayor Daley's televised show. Have we arrived at an age when being well-informed is one with being "out of the action!" Are we moving into a world in which action itself is changing its meaning!

It would seem rather that the very concepts of "in" and "out" are being transformed by electronic technology which has also altered those of "before" and "after." Two skilled amateurs watching two chess champions in a tournament can discover the mistakes of the losing player "after" the game is over. They could not have done so while the game was being played because in that case they would not have been skilled amateurs but champions in their own right. Their hindsight is a kind of wit that comes into its own only when the action is over. The situation shifts somewhat when it has to do with an armchair strategist, a military historian, who knows precisely that the Germans lost the First Battle of the Marne because of the weakening of Von Kluck's army on the right flank. Composed in the tranquility of peaceful scholarship, he has before him all the pawns that went into the battle. In reconstructing it, he can undo in imagination what was done in reality. The historian's hindsight is not a lack of wit which makes up for its slowness after the fact as in the case of the chess amateur. It is the result of his knowing facts pertaining to both armies not then known by the antagonists. The historian is truly "in" but only in the sense of "after." For this reason he is "out." No historian of wars ever won one. Now assume a third case of a man for whom all the pertinent information concerning a war is present to him *simultaneously* in the "now" and even in the "before," thanks to the new computer technology. Assume that he is not an "actor" in that war in any traditional sense of the term. Precisely because he is not actively engaged, he participates more fully in the war than do the soldiers and generals, the diplomats and politicians. And this knowledge gives him, granted the proper existential conditions, *power* over the situation, provided— and here is the catch— that he can synthesize that information, exercise intelligent judgment upon it, and act accordingly.

The possibility inherent fully in the new technology is even today partially actualized. The full use of computer technology plus television would render voting obsolete because predictable results made in the morning would paralyze millions into staying at home and away from the polls in the afternoon. The future of fragmented mechanical democracy today depends upon its willingness to muzzle electronic media. All of this is dizzying to the human spirit because it promises the most radical shift in human values known to history. Power hitherto has grown out of action extending from top echelon down to grass roots. But if power in the future belongs to the observer rather than to the actor, then what was formerly action shifts to what was once contemplation. Decisions will be made by those who *watch,* by the contemplator, and not by those who *do.* Marx's famous critique of Hegel—the philospher observes the passing scene and does not change it—belongs to a world as distant from tomorrow as is the nine-teenth century. This tomorrow implies a return to the tradition of Plato and Aristotle and possibly of Aquinas, but with this difference: they saw a sharp dichotomy between contemplation and action. The philosopher did not hold the sword and this was a tragedy for classical thought. Socrates was the unarmed prophet murdered by the Assem-bly. But if power in the future belongs to philosophers, to those who can synthesize into unity a vision from which will emerge judgments that are fully human, who can shade observed information into phi-losphical penetration and control, then power and contemplative thought will contract their first true wedding in centuries. The sword would pass from the Assembly to Socrates. The activist down on the street would simply be an analytic moment within a total synthesis he can never himself master simply because he will be "in it," not "out." The Marthas are going out of business and the Marys are coming into their own.

There are alarming signs that mechanical culture, commencing with the Renaissance and ending today, is working the tragedy of its death agony back into the heart of the humanity that brought it into being. The creator thus partakes in the death of his creature. He does so in the two principal dimensions known to human existence: the re-ligious and the sexual. The extreme mechanical activism in our churches is like a ride on a merry-go-round without the happy finale when dizzied youngsters rejoin parents on solid earth. The new breed, i.e., the old guard insisting upon the frenetic need to get "in the ac-tion," to be where Jesus is in the inner city and elsewhere, are exhaust-ing themselves in a progressive boredom which settles like mid-afternoon sleep over religious practice in America. Religion in this na-tion is dying along with machines. Millions have by now tired of all the noise. The constant "go-go" seems to get no farther than that Rube Goldberg contraption. Mechanical action that masks meaning in the ecclesiastical order parallels process that aspires to the dignity of being in the technical. The one can pull it off no more than the other. It is interesting to note that the older Latin Mass, now in disre-pute in Catholic circles, is an instance of what McLuhan calls "low definition." This is more compatible with "cool" electronic media than "high definition" services which, in demanding a maximum of physical participation, are better suited to the "hot media" of obsoles-cent machine culture. Liturgical experiments seeking an immediate measurable response from churchgoers treat them as though they were univocal machines that respond uniformly to external manipulation. This kind of experimentation is historically doomed, as is the mechanical society which made it possible.

Religion permeated by mechanical values tends to measure quantitatively success or failure by quantitatively verifiable results: i.e., the bishop who builds the most schools is considered to be a most successful bishop indeed; the layman who can raise the most money is a candidate for election to heaven. Lately, the clergyman engaged in the greatest number of social-action activities has been considered the better clergyman. The number of conversions reported after every Billy Graham Crusade or the number of weekly communions at Notre Dame, tabulated by machine and posted as competitive results, is evidence of success. Whereas the old breed tended to identify the Roman Catholic with no meat on Friday and Mass on Sunday, the new breed identifies him with the regularity of picket attendance and inner-city activity. In all instances American religiosity has been almost neurotically anxious to demonstrate its existence by results that can be verified arithmetically. This possibly explains why the concept of the religious hero as one who stands above or against the world has been conspicuously absent from the American religious scene. Salvation is charted like a graduate-record examination. The highest scores bespeak the greatest felicity in the kingdom of God.

And this moves us to sex. Mechanical articulation of man's response to the real is always visual, as visual as mechanical models themselves. In one sense the "go-go" and the "yeh-yeh" generation points to the future because it defies a rationalist world. But in another sense this same generation looks back to the past because it wraps itself in an activism that is meaningful only in terms of the internal movements of machines. This is revealed in the jerky motions of pale "go-go" girls, who go—nowhere! They dance with men they do not want to touch, and this befits living clocks fearing the invading hands of bumbling children. Wander through a topless joint in San Francisco. The waitresses are curiously boyish or fantastically improbable. But nowhere will you find life. Sexual monsters with Frankenstein bosoms needled mechanically and medically into balloon proportions are not to be touched. They might explode into a dozen pieces as do all decent balloons pricked at a county fair. But we are not attending a county fair. We are in a very old city resisting the future. We are sharing the religious rites of an aging goddess. We move along Columbus Avenue from club to club and we observe this final tribute to the mechanization of the psyche. Objectification of the real permits "The Other" to exist as a mechanical partner upon the distant screen of the visual consciousness, a movie as cryptic and remote as the instrument panel on a B-52. The music booms "My Baby Does the Hanky Panky" but hanky-panky is precisely what will get you thrown out of the joint if you try to pull it off! Thus it is that male and female lose themselves on dance floors in solipsistic admiration as their attention is riveted upon the mechanical rhythms of their own bodies, obedient to orders received from the mummified crown room of the Mechanical Bride on her toppling throne. Not only is "The Other" objectified and thus rendered abstract and distant, but worshippers in the cult are lost in the delight that Narcissus takes in gesturing reverently to the only god he knows —The Machine. These places, full of the sly treason to tomorrow which is yesterday's vengeance, are frequently sought out in odd moments, especially during conventions when anonymity comes into its own, by executives, engineers, and salesmen trained in the new order of things. They mingle with blue-collared factory hands, the "steadies," innocently seeking ocular mechanical solace from female machines, music invading every ear as pounding pistons shake the sensibility in factories. The custodians of tomorrow want yesterday. They are out for a night on the town. They are escaping the future. Their spree suggests the lost adolescence of middle-aged guardians of tomorrow who remember how they once disassembled old Fords in fact as they did young broads in fancy. The discotheque of today blends with those distant years when the old boys read *Popular Mechanics.* They have come back to the tenderloin to find the parts. The harbingers of the new age seek the sweet nostalgia that men know when they re-enter their past. Some like it hot and some like it cool. Even those who work today with computers and transistors towards a very cool tomorrow return to the old heat of yesterday, at least in visual imagination, because it warms up—do not all machines warm?—sensibilities that are jerked back to a simulacrum of life when life mimics, as might a mask, the cold categorical imperative of Our Holy Mother, The Machine. Discotheques join in a final liturgical dirge chanted to mechanics. How-to-do-it sex manuals, mental masturbation by tired millionaires castrated by mathematics, fornication by eyes peeping from dead bodies, salvation through guitars strung from slim hips by boys with empty eyes—they mingle and form an underworld culture as remote from tomorrow's reality as is snake charming in India and peep-hole pornography in shooting galleries on Broadway. The sex machine, like the political machine and the religious machine, is dying as the machine itself dies. The ceme-

tery of the inner city awaits the body. The mourners, somewhat dis-
creetly, will come from suburbia.

Time | Changing Concepts of
And What It Has Done to Us
The "University" and Mechanical Time

A LIFE without the seasonable repetition of ritual fails to chain time and to chasten, if but ever so slightly, the inevitability of death and the cruelty of change. But organic time, in all its rich and voluptuous rhythm, has nothing to do with clock time, dead time, the time ushered into the West with rationalsim in the sixteenth and seventeenth centuries. With Descartes the qualitative dimension of the real drops out of existence and only the skeleton of quantity remains. Quantity is now structured around a handful of iron mechanical laws deduced from necessities of the human mind that see to it that the world "runs on time." As Friedrich Georg Juenger has insisted, the clock symbolized the new world of mechanical necessity. Even the last truly medieval Emperor of the West, Charles V, spent his final days at the monastery of Yuste hurrying from clock to clock, setting their hands backwards and forwards, seeing to it that they ran on time. In an organic order of things there are no sharp "points" dotting a trajectory. Work is suffused with play, the economic with the artistic, the religious with the profane. A continuity between nature and man, God and the World, synthesizes existence and thus weaves it into a unity that can never be dissected analytically into its "diverse" moments. But the very meaning of the "moment" itself can be said to have begun with the time-piece; just as the clock, ticking away, divides time into seconds and hours, so too did the mind from whence proceeded the clock divide life into discrete "moments."

But if time be structured in this fashion then it must be conceived as a Newtonian trajectory spread over the background of space. Time thus becomes something to be "filled up" and to be used. The modern world from the Renaissance until today has been filled with "times" for doing things: there is a time for work and a time for recreation and a time for eating. Where this clock time has been totally divorced from the organic time of man's physical and psychic being, then he is forced to work when his body demands that he sleep: i.e., between one and three in the afternoon. He is forced to eat when his body is utterly uninterested in food: i.e., between five-thirty and six in the evening. He must go to bed when his body finally does want food: i.e., at ten in the evening. This violation has given rise to refrigerator raiding and to Dagwood sandwiches within the United States. It has also ruined otherwise splendid university courses scheduled at three in the afternoon when no self-respecting body wishes to do anything other than rest. Working and playing and eating and sleeping, within a rationalist framework, are no longer themselves temporal events, filled with the sap of the organic. They are nothing more than so many bits of abstract process stuffed into a time conceived as though it were a space. Thus some companies today even count vacations by the hour rather than by the week. You can take two hours of your vacation this afternoon in order to escort your wife to the matinee! Time is constantly filled as life is fragmented into time sequences, each one sharply differentiated from all the rest.

26

At the very dawn of rationalism time became a line leading "out there" to an eternity beyond all time. Thus time struggled to catch up to eternity as did John Bunyan in *Pilgrim's Progress* . Today clubs and other churches that struggle "to catch up" with the age are the last fag ends of this dying rationalism. In the dawn of rationalism, however, eternity was thought to be the inexorable enemy of time, and eternity judged all men "before hand," without taking into account their merits or demerits. Strict Calvinist predestinarianism mechanized heaven because men were saved and damned with the same mechanical exactitude that dominated Descartes' world and that even today makes clocks go round. A rationalist world is governed by the repetitions of quantified motions which can be *predicted;* it follows that human destinies, conceived in terms of clock time, can be foreseen: they are already there "in the works" before the machine is set in motion. Why do little boys delight in springing a watch and thus seeing all the parts go zang? Because they know in some sly and secret way that the watch, like Humpty Dumpty, cannot be put back together again. Thus they take their revenge on a world governed by the strict laws of the machine.

Rationalist time is dependent, of course, on a Newtonian world which sees space as something objective, "out there," infinite, in total independence of the human sensibility and intelligence. Now an objective space must be "filled up" by things and processes that can be seen by the naked eye. This filling up of space "takes time" and hence the objectifying of space involves the objectifying of time. Time is clock time, dead time, duration through a trajectory, and it has violated human or organic time. You cannot measure with a clock the time that lovers spend in one another's company, nor can you measure the time it takes to "get through" a conversation between friends bent on solving all the problems to which the flesh is heir. Such a time is an *intensity*, and when quantified and measured it frequently is in conflict with time understood as duration. Lovers and friends are notoriously unpunctual. Rationalist time thus becomes a commodity that is spent like money. A teacher may give fifteen minutes of his time to a student. Thus the coffee break is a pause, marked by a clock, in the nine to five routine of the secretary. Leisure, considered by the tradition to be a high act of contemplation, transcending or at least cutting through the world of pure work, has been debased to "spare time." The word "spare" is itself significant: a lump of time is spared from work but it is pressed into the service of work because it is "recreation" that refreshes the body and the mind for—more time spent working. Perhaps this objective commodity filling a space reaches its height in the sale the psychiatrist makes of his time in the spatial confines of his office: $75 an hour to listen to your troubles! And all consultants, forced by the very dynamism of their profession, do little more than that: sell their time. Pieper points out a residual resistance to this vulgarism of the spirit in the widespread convention in academic circles of giving a lecturer a "stipend" for his services. The "stipend" suggests that there is truly no proportion between qualitative wisdom and quantitative time. The professor is not paid "for his time" but rewarded for his willingness to share truth with his fellow men.

27

The totalitarianism of clock time spills over into the political order where constitutions are built upon the time span given a president in office. The artificiality could not be more glaring: the president might just be coming into his fullness when his term of office ends but constitutional mechanics demand that he step down. Among the more glaring instances of the imposition of mechanical and mathematical time on the organic is the periodizing of history into centuries, as though celestial considerations governed the hearts of men. Confusing to the student who is thus invited to package the past neatly, the century mark often truly misses the mark. Great changes in history are registered by being felt like waves lapping against the spirit. They come upon mankind very often not in terms of time breaks but in terms of an implicit new life-style played out within the depths of a generation before it finally bursts into explicit formulation. Romano Guardini, in *The End of the Modern World* , insisted that modernity died with the First World War. But even today, more than forty years later, men still speak of "Modern Man" utterly in innocence of the truth that his soul lies moldering in the grave but his body goes marching on.

The concepts of progress and reaction take on new meaning within the context of mechanical time. Progress was invented by Augustine who saw it as a slow growth in virtue and grace to be culminated beyond the grave by beatitude. Reaction initially was simply an organic concept expressing the resistance of any living body to the invasion of foreign hosts. But within mechanical time, progress was conceived as an act of leaping "ahead" over a trajectory whereas "reaction" was an act of retracing steps already taken along the same trajectory. In organic terms neither "progress" nor "reaction" makes sense because time is not extended spatially and therefore is not a reality which can be traversed in either direction.

28

The American tourist in Europe with his Baedecker is often a victim of rationalist time, most especially if he be an aesthetic purist. He cannot bear gothic churches in Spain or Austria whose interiors have baroque altars and retablos. He is disturbed with *all aesthetic transition* because he has become accustomed to cataloguing the spirit according to the dictates of abstract fragmentation. Formed at home by models of "pure" romanesque or "pure" gothic, he is offended when Europe fails to shatter itself into these architectural fragments. Even when he is sophisticated enough to know that history does not conform to types or genres, he vaguely wishes that history did. Having visualized a series of Platonic archetypes which, he senses, correspond to diverse historical periods, he is not at home in churches which took a thousand years to come into existence and which reflect, therefore, the very history itself of western man. But a visual understanding of the real truly must insist that reality be bisected or trisected because the human eye cannot "take in" change, progress, tradition. The only change permitted vision is a change in place, a kind of game of musical chairs in which one age exchanges place with another, one artistic species with another. Our sophisticated American will, however, recognize transitions but he sees them as transitions from one visual point to another: he does not *experience* history as non-linear, as being transitional itself. A mark, however, of the genuine historical sensibility is the ability to grasp the jostling together (our figure of speech is a concession) of all the ages within a man. There, within the catalyst of the imagination, historical time is as impure as is existence.

The hippie contempt for work is largely a protest against this total mechanization of life into moments which are filled up by processes. To the hippie, the Calvinist compliment, the ultimate term within its code—"he is a hard worker"—no longer cuts any ice. The hippie has abdicated from the modern world without as yet having been ushered into the post-modern era. The hippie is transitional, but the Marxist is already doomed. Himself a product of the work ethos of nineteenth-century liberalism, the Marxist—in defining man as an economic function and in rejecting the values of leisure and contemplation—spells out his own obsolescence along with that of machine culture. A story is told about the disturbances in De Gaulle's France in 1968: when the student revolution of the New Left which paralyzed that country was finally taken over by the highly-professional Marxist machine, the bourgeoisie in every cafe in Paris breathed a sigh of relief: things were getting back to normal! Bourgeoisie and Marxist define one another. But if the Marxist is doomed in the long run, then so too is the successful business executive who tries to "keep up" with the enormous mass of information fed him today by a computerized and electronic technology. On a treadmill that refuses to let him move ahead and "get on top" of the information, just like a child running down an escalator that is going uphill, the executive had better learn how to relax and watch the information go by. Otherwise he will collapse before history has had a chance to teach him his new role. You can no more *keep up* with electronic technology than you can fit the principle of the wheel into atomic fission. Mass machine production along with the fragmented time that made it possible is today almost as dead as the steamboat. And the hippie is still around to mock those of us who do not see it. He has all the time in the world!

The Greeks symbolized that act which defines—insight—by the word "eureka." I've got it! Older comic strips pulled it off by a little light flashing above the head of someone who just got the point. Insight into the real is never temporal, although it would be rash to call it anti-temporal. Crossing time, emerging as the result of a searching through hitherto unpenetrated materials and of a meditation upon them, insight is preceded by time but itself "inscends" rather than transcends. Because insight inscends it cannot be measured temporally. Intelligence tests simply measure the ability to get through a

29

number of questions during a given period of time. As McLuhan points out we are dealing here with the measurement of things in a series. For example, the Graduate Record Examination, dependent totally on duration, confuses being with duration and thus determines intelligence by the swiftness with which eyes get across a piece of paper. But existence is not duration: it is non-cessation. Bergson almost grasped this when he saw that the fluidity of life, once conceptualized, is frozen into discrete moments which, we might add, can be made to move like the frames making up a motion picture. These frames *look like* life in a fashion analogous to the way in which a Graduate Record Examination score looks like the intelligence. But every wise professor of those humanistic disciplines which touch upon the mystery of being knows that the best way to get at a student's mind is to let him symbolize meaning in his own fashion. Meaning is then seen as bathed in an existent. These symbols are not mechanical. They cannot be counted like arithmetical units. In fact they are usually verbal, thus escaping the visual arrangement of problems upon a piece of paper, a space. Insight does not "move across"; it "plunges into." (Even here we are plagued by the linear structure of our language.) Like Gerard Manley Hopkins' "inscape" insight catches and then marries the inner rhythms of the world.

30

Given that we are leaving a horizontal age and moving into a perpendicular one, our educational system on the university level is hopelessly outdated. It served our civilization well during the ascendency of rationalism but it cannot prepare men for a time when decision-making at leisure will be their main "occupation," although this last may not be prominent within their hierarchy of values. Why should it take exactly four years to get a B.A., two more for a master's degree and then two or three for a doctorate? Why one hundred and twenty-eight hours for a degree? This quantifying of the qualitative belongs to the mechanization of time, its division into years and hours. Similarly the mystique of so much "material to be covered" suggests a mass of information spread out into space which must be gotten through in a time sequence by both teacher and student.

The compartmentalized and duration-oriented universities of the West, beginning roughly with the Renaissance, broke sharply with their medieval predecessors. The medieval university was largely a place where professors and students talked to one another. Things were written down in textbooks simply as a prop for memory. The dryness of the theological and philosophical *summae* reveals their essential notebook character. They served as instruments for memory and for the tongue. Nowhere is this better preserved as a symbol than in the University of Salamanca which guards to this day a classroom dating from the fourteenth century. The wooden benches are carved with hearts pierced by arrows, thus revealing that even ecclesiastical students were not immune to love. This gesture to Venus links us in a common humanity to those dead and distant students. But the eye is riveted upon the professor's pulpit. There is nothing like it in the modern world and it divides us from those ages as would a chasm. Made out of one solid block of wood, this carved masterpiece supports at the base a little bench where a student sat and read from a text; above is the pulpit, high and removed, where the professor commented and added and argued and in general imposed himself upon his audience. The priority of the spoken over the written word could not be better fixed as in a sign for meditation. Given that the intellectual world was quite literally a theater of discourse, the scholastic dispute between masters who fenced at one another before noisy audiences of students who cheered on their respective champions has been lost with the passage of time. The Gutenberg revolution stands between us and the Middle Ages like a wall forever covering as would a silence the jangling color and vivacity of an age that could only be recaptured if past sounds could one day be recovered by post-modern man. The best part of the Middle Ages is simply lost to our highly-literate time because what is not transmitted by being carved in stone or written down on sheets of paper is simply dead to us: oral tradition, even while it lingered on in backwaters and pockets of resistance, died under rationalism's insistence that if it isn't written down, it isn't at all.

It has been suggested that we know more about late Republican Rome than we know about the England of the fifteenth century and the War of the Roses. The former was preserved in the timeless prose of a literate aristocracy. The latter was lost in the clash of steel upon steel of letterless Christian knights. And only hints remain of what the University of Paris was in the thirteenth century: a student could throw a mug of ale at a professor who shirked his duties but he was forbidden recourse to the dagger; fines were meted out in terms of pints of beer; bands of Latin Averroists clashed with Augustinians in narrow and tortured streets on cold winter nights whose sinister shadows promised the poetry of Villon and the *Danse Macabre* , the ultimate winter of an old world.

There was a kind of hurry about medieval life that precluded periodizing the time span of a man. Men who matured rapidly and died young were already teaching masters in their late teens and early

31

twenties. Life was not fragmented into childhood, adolescence, maturity, and old age. Unlike later theologians Aquinas did not fix a "time" when man came to the age of reason and could sin. He contented himself with simply stating that this "time"—whenever it comes—is one with the ability to discriminate right from wrong. Preparation was not for life but was life itself. The great unsung hero of the West, Baldwin IV of Jerusalem—blind, a leper, and a fighting king— assumed an impossible crown at fifteen and was dead at twenty-five. He did not prepare for a short life: he lived it like the man he was. But within all this hurry, to return to the university, all disciplines worked together to form a unity which was a spear aimed at what Augustine had called Christian Wisdom. Liberal Arts served Philosophy and Philosophy served Theology. The medieval university was synthetic rather than analytic in its goals. The *summae* themselves reveal how masters and doctors bombarded their subjects, usually sacral, with a bewildering mass of information that ranged from Greek philosophy through Roman rhetoric to the ruminations of San Isidro of Seville on bees and carpentry. Nothing was pursued for itself and therefore everything was pursued for a contemplative ideal which forged into unity what otherwise would have been a dust of information ready to be blown back into the barbarism of the centuries from out of which the medieval unity had been forged. The synthetic character of medieval intellectuality has been so lost that an entire generation of scholars earlier in our century tried to disengage Aquinas' philosophy from his theology in order to package the two of them neatly. Gilson has insisted that such an enterprise not only violates the mind of Aquinas but that it is an impossibility.

But if the medieval university was synthetic, its modern counterpart, beginning roughly with the Renaissance, was, and has remained until today, analytic. The discovery of printing and the breakup of religious unity removed philosophy from the classroom to the lonely studies of a Descartes, a Spinoza, or even, despite his courtly gregariousness,— of a Leibniz. The spirit spoke to a piece of paper over which hovered a pen. Insights were hammered into print in the solitude. Let us never forget that Descartes feared the wrath of the still basically medieval University of Paris. The fragmentation of reality produced by the Cartesian method which consisted in dissecting clear and distinct ideas and then fitting reality upon this Procrustean bed: the release of a new technology based upon mechanical models which conceived the world in terms of the clock; the gradual shattering of the synthetic unity of life in the name of a new political absolutism that atomized the rich complexity of medieval society as it centralized all power within its own hands; all of this and even more produced a new kind of university that was structurally analytic. Each department enjoyed a sovereignty and dignity untouched by communion with its neighbors. But analytic fragmentation involved the mechanization of time. As disciplines were pared away from one another the very momentum initiated fragmented each discipline into segments, conceived as though they were parts of a trajectory. Since it takes time to pass through parts on a linear line, it soon took set blocks of time to pass through the segments forming a discipline: hence the rigid time sequences—so many hours in this or that field makes the master; hence the gradual emergence of that peculiarly modern animal known as the "scholar." The research scholar is not marked by his ability to penetrate a subject matter with insights; he is not acclaimed a scholar by his being at home within "the order of judgment" (to use Aristotle's language) in which he structures into unity, and thus synthesizes a body of knowledge. John Henry Newman tried to do this at Oriel and was flunked outright for not knowing a strictly-defined body of information found within a narrowly-enumerated list of books. The scholar is marked, rather, by his *expertise* within "the order of discovery" (again, Aristotle), by his capacity to answer the four analytic questions: who, when, where, and what. The scholar thus is a man who discovers "what is" and "how it came to be" within a well-defined and fragmented discipline that jealously keeps out any competition. Super detective, the scholar truly came into the fullness of his perfection in that same nineteenth century that gave birth to the detective novel. Sherlock Holmes and Alfred Harnack both detected within disciplines whose boundaries were as well marked as are those of a French formal garden.

33

The scholar has been ridiculed in song and story. His file cards and fat libraries, his pedantries, his crusty pride, his propensity to learn more and more about less and less have made him a natural target for both populace and poet due to their mutual discomfort within the compartmentalized world of the university. The butt of popular wisdom, the scholar has nonetheless been the chief spur and impetus of our civilization since the Renaissance: neither medicine nor the physical sciences, neither the enormous advance in sheer humanistic information nor in history would have been possible without the analytic mind and the work mystique that dominated the university.

But the traditional research scholar, as well as the university he made possible, is today passing rapidly into history. A new space-time ratio, forced into existence by the new electronic technology as well as by the simultaneous coming into perfection of the older and antagonistic technology, has created a crisis within the academy. As Rollo May points out, the widespread student revolt, whatever its political motivations may be, reflects dissatisfaction with an institution that has swollen to such immense proportions that it resembles a factory bent on serving the community the commodities it demands. Initially a blend of the older contemplative ideal (now secularized) and the new fragmentation, the scholar served the community by serving himself and by following the bent of his own interests. Society picked up his research and did with it what it wanted. The scholar was left at peace—hence, the Ivory Tower. But, as Juenger insists, technology's perfection is its very failure. The stunning success of the western university has so eliminated the disinterested and the contemplative that both teacher and student feel oppressed and *used* by a community they often reject.

Nonetheless, this moral crisis faced by the university pales before the crisis produced by electronic simultaneous information. This crisis renders obsolete old-fashioned university space and time. If we wanted to we could hear the best of the Sorbonne, of Cambridge, Freiburg, or even Leningrad in every college in the United States, beamed in by Early Bird satellite. With simultaneous translation every language in space and time becomes every other language. Something will be lost: the richness of the original; something will be gained: the living image of a master. Under the pressure of this simultaneity of information only the very top professors will survive. Ironically enough, in our rush for universal education we may find that we need far fewer teachers rather than more. We may discover that the entire educational establishment, today formed of hundreds of thousands of teachers, will shrivel to a small elite of intellectual and rhetorical geniuses who will flash their message to the whole world.

The research-oriented scholars, being turned out today by the tens of thousands, are analogous to horses kept in reserve by those daring souls who bought the first automobiles. The car in the front yard and the horse in the back blended daring with prudence. Soon enough prudence was proven hindsight. Research scholarship, as we have suggested, is principally an affair of fragmented time sequences—getting through materials in weeks, months, and years. But with information *simultaneously* at hand thanks to computerized technology, there will be no way, quite literally, to measure the time needed when a man is fit for a degree. The very concept of the degree itself will change. The sweating will have gone out of the university business and nobody will get prizes or encomiums for doing work already done by computers. (The shock to a morality geared around the work mystique will be enormous.) Intellectual competence will be judged in terms of the ability of the student to synthesize the explosion of information. Most significant thinking will be reflective. With Aristotle's "order of invention" absorbed by computers the need for intellectual detectives will be radically reduced. Men will succeed or not in the measure of their ability to order information into unity and to evaluate and judge (Aristotle's order of judgment again, his very principle for distinguishing wisdom from mere science).

It follows that all higher cognition will be philosophical or will partake of the philosophical. This will render obsolete most of the philosophy departments in the United States. These last are either historically- or positivistically-oriented. But both historical and logical analysis will be absorbed by the computer. The paperback edition of Kaufmann's *The Age of Analysis* has a cover depicting a bald and grim young man who has been ordered, one presumes, to think by the numbers. But thinking, sequential thinking, involving—as it does—the analytic dissection of the given into its parts, is obsolete. Linguistic analysis itself, the darling of a discipline which has been duped into aping the exact sciences, can be done far better by computers than by man. The philosopher of the future will not have to think: he will have to judge: this is a horse of a different color.

What we do with all this new information depends entirely upon the transcendent question of the "why" of the information. Programming will be philosophical because we will have to decide *what* is significant in the light of *why* it is significant. Programming will be determined by answering the question of "why the being of this" rather than "the being of that." Why the Visigothic code rather than Viking runes in Minnesota? Why the jota of Navarre rather than the flamenco of Andalusia? Why the Apollo space program rather than cancer research? These questions and the decisions issuing from them will lead mankind again into the ultimate and sundering question: why being rather than non-being? This forcing of the race into a new awareness of contingency, of the chandelier character of the planet itself at the hands of man, into the area of what Paul Tillich called "ultimate concern," will carry with it the frightening possibility of an ultimate anxiety gripping an entire world. We will literally have to grow up in the face of an age which will have done all our homework for us. The only alternative to the psychiatrist's couch will be a new Age of Metaphysics.

Meditative reflection upon information already possessed by the intelligence points towards leisure. The student of time is therefore invited to explore time's relations to leisure and to reflection. Both bespeak time's relation to hierarchy. All reflection preceding decision and judgment "takes time" but this time is organic. It is not mechanical. There is no "measurement" involved—only a qualitative intensity which is experienced by a superior mind when it sweeps inferior judgments into itself, thinks them through, absorbs them, and then transcends them. Bernard Lonergan is right when he states that the "upper level" judgments of better minds permit them to understand their inferiors whereas the converse is impossible. This impossibility that dimwits understand bright-wits is a capsule definition of hierarchy within a world that does not recognize any external ultimate Authority in matters relating to Truth and that has also rejected the rationalist insistence of absorbing countable numbers under univocal types. Within the rationalist theatre of discourse hierarchies are established by testing men according to some objective standard, some *tertium quid* . This last always turns out to be arithmetical: so many bishops favor this course of religious action and so many politicians that course of social action: all of them are authorities! Thus intensity of applause, itself the result of a superior number of hands clapped together, indicated victory in intellectual debate just as political probity is always, as the term goes, "tested" by the electorate. The analogy with Gallup polls for entertainment figures in the mass media could not be more obvious. The withering away of these quantitative standards indicates that no public Authority is likely to be recognized as such in the immediate future. But external measurements based upon counting will be impossible within the coming age of reflection.

Hierarchy will be determined internally by a man's *knowing* his own insights as well as those of men less bright than himself. The only "standard" will be this: in an intellectual debate which affects the whole community, the true philosopher will be the man who can think through his opponent's arguments, even use them ironically and rhetorically in order to relieve his own irritation at having to face inferiority, skim off what is viable in those arguments, and then transcend them. His inferior, for lack of "upper level" judgments, will never be able to do the same when faced with more profound insights than his own. He will simply not be able to *understand* what his betters are talking about! Hierarchy, therefore, will be interiorized until—should this happen—some new public Authority emerges. Such a public Authority, being brand new in our world, would have nothing to do with

36

the old mystique of counting, of "public opinion," of the egalitarian aping of a uniform and machine technology by then dead and buried except in the most backward of countries.

The very recognition of intellectual superiority and therefore of hierarchy will depend upon a resurgence of a very old virtue, humility, a humility that surges up from within the inner recesses of the spirit and that renders it possible for a man to locate himself within a new social structure altogether bereft of "spatial" or rationalist "localization." That Aristotle himself insisted that such a humility is a rare gift, itself a sign of *moral* superiority in the man possessing it, renders the situation even more ambiguous. Nonetheless, student mobs that today can intimidate professors who are spatially present to them will have a hard job in doing the same with professors who beam their lectures into Berkeley from some distant mountain retreat in Ecuador. This moment of transition will soon be transcended by the West. The public confusion and embarrassment produced by a situation in which there are outmoded mechanical criteria for excellence will soon pass. It will give way to a new interiorized awareness of competence and superiority based on reading another man who cannot read you. This will be difficult and painful. It will be open to sheer intellectual fraud. The transition itself will be rendered somewhat easier to bear thanks to the destruction of the older rationalist patterns still largely dominant today, rationalist patterns which permit sheer weight of numbers and the violence implicit within them to overawe scholars who by their very nature are not inclined towards heroic resistance.

37

Reflection cannot be measured in terms of mechanical time. An act that is thoroughly qualitative can only be "weighed" by the spirit. The time it takes to come up with an insight in any order of reality is often totally disproportionate to the insight itself. Yves Simon used to say that philosophical "projects" could never be measured in terms of the time element involved in bringing them to fruition. A grant of a year's duration given a man to explore some subject might be unrelated to any "achievement" wrought therein: the insight of an hour's meditation or of a meditation prolonged through years of study simply cannot be structured within foundation grants that are based upon a rationalist division of time. We do not truly care how long it took Shakespeare to write *Romeo and Juliet* or for Bach to compose *The Art of the Fugue* . Artists today bent upon being subsidized and foundations bent upon subsidizing them have entered into an uneasy alliance. The foundations must dispense their funds according to rationalist time sequences, so many dollars for so many months spent on "research and/or creative production." Frequently they lavish young men and women with cash totally out of proportion to the clock time they need in order to bring their work to fruition. Very sensibly these young artists pocket the money. The foundations, well aware of what is happening, simply look the other way. Both artist and collective patron know implicitly that the time is soon coming when pay will no longer be measured by time at all. The loafing permitted thousands of young bearded chaps who officially are "researching" Navaho pottery and Aztec snake symbolism is simply a bow proffered a future which is almost upon us. This future, of course, is a presence in both medicine and management consultancy wherein men are paid in proportion to the *intensive* value of what they do. It seems ironic that the artist must still labor under the very mechanical structuring of time against which he, almost singlehandedly, led the reaction in the full flush of romanticism in the last century. We need only think of Whistler's taunt, thrown out under oath and in court, that he was not paid two thousand guineas for the twenty-four hours it took him to complete a painting but for "the knowledge of a lifetime." Admiring the splendid arrogance of Whistler's remark, we must see it nonetheless as a backlash within a world whose rhythms were determined by the clock. Why did Whistler have to excuse himself on temporal grounds? Why did he not boast that he just dashed off his painting without previous preparation? The point is grasped when we see that excellence and worth, intrinsic value, will no longer be in proportion to clock time. Neither the brilliant razor of genius which hits it off in the flash of an instant, nor a work that takes a lifetime to mature into being, will matter at all. Mechanical time will be transcended in a fashion that parallels the simultaneity of computerized technology. Just as the scholar will be judged by what he does with information fed him instantaneously, so too will the artist, as well as the scientist and the manager, win his spurs by the internal worth, the being, of what he brings into existence. Time will no longer be a factor that pays *either way* . The very concept of the sluggard will pass out of history. That artists and quasi-artists and would-be artists flee the United States today in order to hide in Mexican villages untouched by mechanized time bespeaks a mysterious relationship between organic time and personal creation. These latter-day Bohemians are judged, not on what they can come up with on a nine-to-five schedule, but on what they come up with, period!

A slow-down in a factory reduces the end-product because this last is purely quantitative, but a slow-down in musical composition or

38

philosophical meditation cannot intrude upon the value of what artist or philosopher does. Traditionally these acts, all of which belong to the order of "insight," pertain to leisure because they cut across those processes in which man exhausts himself in the other, in what Yves Simon called "the generosity of labour." "Leisure time" is not true "leisure" because the former exists for the sake of work and is a "re-creation" of the body and of the mind for tasks that still lie ahead. "Leisure time," therefore, belongs to the world of work. Even Marxists who are sworn enemies of both disinterested contemplation and play see to it that socialist workers get sufficient recreation. "Leisure time," however, will soon be obsolete for "decision-makers" because the sharp distinction between a time for work and a time for leisure will be blurred in an age in which most significant activity will be one with reflective thinking. Men will not be " *in* leisure" as though leisure were another space filled by a time sharply measured by a clock.

Men will be " *at* leisure" most of the time when they are working. The very ratios by which we have measured time will be reversed. Until now mankind, or at least the male half of western mankind, took care of domestic and personal chores during whatever "free time" they had from work. In the future what was once "free time" will be work and what was work will be leisure. The old lines, today obscured, will be turned topsy-turvy in an age in which what was called work will be thoroughly interiorized by having become decision making. To be *at* leisure will be man's work. In fact the very terminology will be obliterated by a radically new order of things. Men who will not be able to stand all this leisure, who will fill it with banalities, will cave in under the pressure. Fragmented leisure is self-defeating and a lifetime of superficial play is exhausting. Even now darkest suburbia experiences great exhaustion produced by its inability to handle invading leisure.

39

Barring some atomic catastrophe which would presumably reduce our society to a simple way of life, we can trace today the mainlines of tomorrow's world *in thought*. We cannot *visualize* with any accuracy what that world will "look like." We must surrender such visualizations to science fiction writers. We do know the following: work processes will largely be done automatically and electronically. Even man's "work time" which was his old "free time" will be drastically curtailed. Bill-paying and other first-of-the-month nuisances will be absorbed by the new technology. Handling of money will be a thing of the past. Grocery shopping will be done over monitored television if there is a grocery store. Even today robots are available for housework but they are still too expensive even for the rich.

All of this points to a shift in space as well as in time. Spatial *ne-cessity* will disappear as a significant factor in life. We tend to project the future in terms of the present, and therefore Jules Verne's twenti-eth century looks more like a speeded-up Victorianism than the world we actually know. We are tempted to think of space as being even more filled with instruments of technology than today. The facts sug-gest the opposite. Technology will be less obtrusive than we now expe-rience it. Machines will get smaller, not bigger. The entire spatial structure of industry, most especially office organization, will be taken over by computers, thus breaking up the highly centralized office buildings of today. This decentralization will obliterate fixed space in which men make decisions. Man's being *at* rather than *in* leisure will jell with a new space. Executives as well as educators will be able to work wherever they please. They will be able to monitor their deci-sions from any place they care to—from home, from boat, from rural retreat, from anywhere in the world. The mass exodus from the office will set some men to travelling indefinitely and it will return others to domesticity. They simply will not be able "to go to work." There will be no place to go.

Every man will be "connected" with every other man. The older privacy was a necessary defense mechanism against the compartmen-talized fragmentation of machine technology and the mentality that created it. Privacy, as we have known it, defended a man's differences from a world that placed a premium on sameness, on monolithic uni-vocity. You hid your differences as precious idiosyncracies threatened by an order that demanded machine-like conformity. Differences, cherished as personal and domestic values, were simply monkey wrenches if thrown into the clock-like precision of a world of ma-chines. This sharp dichotomy belongs to the dying rationalism of the end of an age. Tomorrow human beings will be defined (Does not the very verb suggest an irony if not a paradox?) by their differences. The univocal sameness in mechanical repeatability that goes with Carte-sian fragmentation will give way to an analogical world in which men will be forced to live up to themselves and to their own possibilities rather than adjusting both to the grey-flannel standards of company organization. The company will not even be seen because it will not be "there" as a visual object filling a space. A good detective novel might be called: "Where is the Company?" The Willy Loman of the future will be the Company itself. The entire social world centered around the "organization"—company wives and convention smashes and the whole bit—will wither along with its *raison d'etre*. Since only quantitative and hence repeatable essences, the analytic order, can be measured—so many Fords, all alike; so many executive secretaries, all alike; so many beauty queens, all alike—measurement of individuals against univocal standards will be rendered obsolescent. The age of abstract essences—from the good guys to the good Fords—is going out of business.

The very symbols by which society expresses itself will be non-visual, as non-visual as is the Thomistic analogy of proper proportionality according to McLuhan. The new society will be structured in such a fashion that the only philosophical analogue we can find for it within the tradition of the West will be that of the same analogy of proper proportionality: every being is absolutely different from every other being but relatively alike in that each being is related proportionately to its own existence as is every other being; likewise every man will be absolutely different from every other man and only proportionately alike in that each man "measures up" or fails to measure up to possibilities uniquely his own. We see hints of the new order today: fewer women, especially young women, try to dress glamorously in order that they might measure up to a common standard of beauty.

41

Conventional science fiction has generally missed the contour of the world being shaped around us. Projecting de-humanization into the future science fiction paints a grim and ghastly world peopled by machines that oppress the human spirit. It depicts robot-like men marching to the inhuman totalitarian rhythms of a society from which organic life has fled. Actually these jeremiahs, high priests of the intensely moralistic genre of science fiction, are describing the asymptote of the world of rationalist machine technology. Their dire predictions may contain more than a germ of truth because, as Juenger predicted, technology reaches its perfection in the very moment of its immanent death. But the new order, fashioned around electronic technology which is rapidly replacing machine technology even as this last hammers out its formidable finale in history, will not be an impersonal world. It will rather be intensely, and possibly intolerably, personal because of the new time-space ratios described. So intensely personal will be the new order of things that we can wonder whether men, once released from the need of being measurable objects, living up to school or company or suburbia, have it in them to live up to their own liberated personalities, have it in them to become subjects of existence and centers of responsible liberty.

James Jeans, attempting to popularize Einstein's world of relative time and curved finite space, once insisted that the concept of "relative simultaneity" had its existential counterpart in the real world. Therefore, he wrote, if you are in the death cell and about to be executed this morning, take comfort! From the standpoint of an observer on another planet, you are already dead! This highly-involuted consideration can hardly give comfort to a man who knows with a deadly certitude that he is now alive and that very soon he will be dead, every scientific theory to the contrary. "Relative simultaneity" makes no more existential sense than does the insistence of sophomores studying physics who pontificate that our senses deceive us into believing that solid oak is really only a whirling world of atoms and neutrons. They confuse, as did Jeans, a convenient mathematical construct which symbolizes reality with the reality symbolized. Thus symbol becomes thing and thing becomes symbol. Men of such a bent soon come to live in a weird Alice in Wonderland world in which reality is abstraction and abstraction reality. One is reminded of Greta Garbo's "Ninotchka" in which she plays a Communist intellectual who insists that her having fallen in love can be measured in chemical and mathematical terms, thus charting rationally the rise and fall of her heart. But the symbolic interpretation man gives his world ought not to move from the abstracted intelligence to the real but from the artifacts produced by the abstracted mind *back* to the human sensibility. This sensibility is profoundly affected by the cultural and artifactual world thrown into existence by the mind. This new world, the product of the human spirit, now invades the psyche and fashions it according to new sense ratios. In Brooklyn who ever saw, *literally* saw, a tree grow? But in Brooklyn who has not seen, as a child, a world of asphalt and concrete and of streets and pavements rise into existence? And who, then in Brooklyn, was not profoundly affected in the depths of his soul by what he saw?

"Simultaneity" can never be "relative" in any temporal sense which would contradict time's "before" and "after" but "simultaneity" can be relational in the sense that it instantaneously relates time sequences. Electronic simultaneity of information which forbids dissection into fragmented analytic moments which man must "cross" or "get through" will eliminate, as suggested, the scholarship in the sense of pure historiography. But here we must make a distinction. Kant's complaint against Cartesianism was based on his insistence that pure rationalist analysis can never *add* to our knowledge. Analytic thinking can simply penetrate and arrange the already given, the already possessed. An archetype of analytic thinking is the butterfly chaser bending over his prey, dissecting it into wings and body and legs. He cannot add; he can only count what he has bagged in his net. True progress in knowledge—not specifically in the sense of knowledge possessed by me as an individual, but in the sense of adding to the entire store of knowledge possessed by mankind—is synthetic. This is well expressed in Kant's discussion of the synthetic judgment in which he demonstrated that the predicate is not analytically implicated in the subject any more than red hair is analytically implicated in the concept of man. Applied to the discussion at hand this means that historical discovery, the synthetic addition to what is already known, cannot be fed back out of a computer. The computer can feed back only what is put into it. The work of the computer is thoroughly analytic. Not only does it do this work much better than men can do it, but it does it simultaneously, whereas analytic thinking by a man involves sweating it out through time's duration—ask any graduate student in history! The discovery of the already known by the individual to whom it is not yet known has formed the discipline of historiography, the spine of modern historical scholarship. The teacher does not tell the student what he knows because the student must get down the discipline, learn his trade. But this trade is now obsolete even though its practitioners, understandably enough, are reluctant to admit the fact.

43

These considerations will radically transform the study of history. Historiography will perforce give way to the historical imagination because this imagination truly does grasp history *simultaneously,* forming thus a human analogue to the electronic revolution. Historical time, annealed within the catalyst of the imagination, becomes time past contained in time present and time present in time past. Time, an analogue as well of eternity, ceases to be a parallelism running backwards into a dim and distant "past." History becomes contemporaneity and the possession of the historical imagination can be judged in terms of a man's ability to *be* not only within, but of, an age now past. To know that Charlemagne was crowned Roman Emperor on Christmas Day in the year 800 A.D. in Rome is one thing but to be there with Charlemagne being crowned is something altogether different. This much Collingwood teaches us. But he wrote at a time when historiography was at war with the historical imagination which it dismissed as pure poetry or "subjective history." Certainly the past has no "objective" existence in any temporal or extramental sense of the term. The past has been but is not now. We can only leap back if we conceive history to be a linear trajectory occupying an infinite space.

For this reason men first decided to take a jump backwards into time during that very Renaissance that gave birth to printing, itself the effect of mechanical thinking. Both the humanist attempt to return to the cool gardens of antiquity by skipping merrily over the messy Middle Ages as well as the reformers' desire to reverse time by traversing the ages backwards—thus walking down a trajectory "the other way," in order to reach the pristine doctrinal and liturgical purity of the Gospels and of the Apostolic Age—were consequences of a new space-time ratio. If we drop this rationalist linear time, history becomes just as contemporary as is the man who knows it. Quite literally, that man *is history*. History's being is his own.

Marshall McLuhan has suggested that medieval manuscript culture understood light not as something which was "turned on" but which "shone through." Light, itself a symbol for both intelligence and eternity, bespoke an eternity which shone through time rather than ran parallel with it. Analogously, the shining through of historical time upon a sensibility bathed in an intelligence attuned to the historical order does not produce but *is* a simultaneity permitting man to manipulate history at will. Within the historical imagination lost causes can be retrieved and victories can be denied: do not all Englishmen today divide ultimately on their sentiments concerning a civil war that was fought some four hundred years ago? The imagination manipulates time in such a fashion that even literary genres find their fulfillment in history. History becomes the hub of the wheel of education, the forge hammering into imaginative unity not only the literary but the philosophical and theological traditions of the West. Even the scientist's reflection upon his own disciplines and his coming of age as a philosopher of science will be exercised upon historical materials. Insights will be philosophical but they will play over a vast body of historical data fed the mind by a disinterested technology in the service of man. The synthetic structure of a life hitherto fragmented into analytic departments was the province, in the past, of a limited number of talented cranks. We need only think of a John Randolph of Roanoke or of a Winston Churchill. Neither would have been invited to occupy university chairs because their very eclectic vision offended the compartmentalized mind. But the knack of synthesizing in the future will be the heart of education itself, this of course on the assumption that education has a future.

St. Augustine insisted that the entire past makes sense in terms of a middle, looking back upon a past which itself leads into a future. In his own theological terms, the Incarnation illuminated what otherwise would have been a meaningless past simultaneously with its pointing man towards Apocalypse and Judgment. Translated into existentialist terms by Martin Heidegger today, this means that my personal past, at any lived moment of human time, *means* precisely what my future means. A carpenter is not likely to remember his eighth-grade literature. A housewife and mother is apt to forget what she learned about Plato's *Republic.* The very dynamism of life forces us to select and choose out of the past that which is of value to our future. Synthesis is always personal and therefore moral, philosophical, in that it involves a stance, a point of view, a preference, a future. I stiffen my past into a structural unity by paring away whatever is irrelevant to the moral thrust of my existence, to where I am going, to my own understanding of the "why" of my life. I forget, and this is a blessing, practically every banal fact that ever happened my way. Who can reconstruct with accuracy every external motion, to say nothing of fleeting phantasy, experienced in one hour's period? There would be something psychologically wrong with the man who could do so. Joyce tried to do precisely this very thing in *Finnegans Wake* where he reconstructs a twenty-four hour period. In so transgressing the grain of human existence he sinned against normality, even though his union of form and content was an artistic precursor of the electronic. We remember, if we are sane, what we want to remember and we flood that memory with the intentional projection of our own future. Man present is always man past and man past is man future. The rupturing of this existential structuring of human life bespeaks either oncoming senility and closing down the future; or it suggests an insanity that sees no meaning in the past because there is none in the future. Show us, then, what an age remembers and we will tell you where it is going. Show us what an educational system considers important in the past and we will tell you its plans for the future. Tell us what a man remembers of his boyhood and we will show you not only where he is going but— and this is the same thing—what kind of man he is.

And if we would see this psychological drama which is each man's very history as it works itself out within *corporate* existence we might well attend to America's understanding of her own past. If we attend very carefully to the meaning of human time as developed in this study, we must conclude that America is in trouble today. America has failed to live out in any normal fashion the structure of human history. Our American understanding of time has always been spatial, not really temporal at all. Time, the past, was something that lay behind our backs as we looked west. We thought we could carve out our

45

future by dropping our past, that space that ended on the far side of the Atlantic. History has meant Europe for America just as history books for our children have always suggested a remoteness, a distant trajectory unrelated to the present. Children over seven find it very difficult to gain an historical sensibility absent in those first crucial years of life. But America's mythology has conspired with psychology to make it almost impossible. We did, of course, have our history which was stiff with battles and marching men, with George Washington crossing the Delaware in a tapestry, with blue- and grey-clad boys at Gettysburg and all the sad magic of flowers on graves come Decoration Day. But this history—unless we were Southerners matured by being annealed in defeat—was as much the furniture of childhood as are parks and picnics and Fourth of July oratory. Our history has always seemed infantile because it has always appealed to children who then grew up only to be taught that America *really* meant the Future. But this cocky confidence which bred Manifest Destiny and Teddy Roosevelt has today died in the holocaust of our own inner cities. It has died in two lost wars which contravened what they told us as children: America never, but never, lost a war! Today it seems that America never, but never, must win one! And therefore, America—not yet fully aware of a tremor which is passing through its soul with a greater inexorability than does the Andreas Fault pass through the body of California—is looking into herself, is searching for a soul which would be her very own. In terms rendered popular by the periodical press, America is passing through an identity crisis.

What are law and order in the streets? What are law and order anyhow? Are they different in black ghettoes than in white suburbs? Who are we anyway? Bereft of any public orthodoxy erected into being from a fixed point of view, America is as frantic as is a woman who just lost her purse. Or, to change the figure of speech, America is a sailing ship in stays, drifting because the afterguard cannot agree upon a course. Until this painfully unsettling problem of a course, of a destiny, of a national definition has somehow been settled, America will drift off a lee shore. America was taught—we were all taught—that this was a young nation. Nobody pointed out to us the simple historical lie. America is an old nation. No nation in the West, excepting England, has had a longer uninterrupted political existence. We tend to think of Austria-Hungary and of Germany and of Italy as being infinitely aged in comparison with ourselves when all the while they were children when we entered into our full maturity. The Germanies passed from Holy Roman Empire to Confederation to German Empire to Weimar Republic to Hitler and then to Bonn and permanent division in a time that does not even equal our life as a nation. France knew kingdom and republic and empire and then kingdom again and then still again a kingdom of another kind, and then France knew republic once more and empire still again and three more republics. All the while the United States remained one in its very ancient being. Spain suffered six devastating civil wars that ravished its land and tore its soul forever asunder, as swords plunged into a parched soil and there became crosses that marked the graveyard of dreams that still haunt the living as unexorcised ghosts. And all the while America suffered but one Civil War that simply confirmed an already-venerable corporate existence. Austria-Hungary—a name that conjures up the Eagles of an Empire as old as Charlemagne—only came into history some seventy years or more after we were already in being; and Austria-Hungary has already been dead some fifty years! Italy was a geographical abstraction when we were flourishing into prosperous middle age. Our own revolution antedated the French Revolution. In all things we are the patriarchs of a confused civilization.

47

America is truly an athletic but worried man on the wrong side of middle age, and this athlete is deeply troubled because he suddenly knows that he remembers everything that happened to him from age twenty to age fifty, but nothing earlier! This aging if muscular gentleman has been taught from early youth that he ought not to remember but this does not help him enjoy his amnesia. Everywhere he looks he sees grinning and knowing faces from out of the void. They mock him with the knowledge of a memory that has fled but that is indecently known to all but himself. He is like a conscientious and honorable drunk on the morning after: what happened to me last night? Did I disgrace myself? His amnesia prevents effective action in the future because amnesia is innocent only when it is not known for what it is— amnesia. Ignorance is only innocence when it is not knowledge. Surrounded by file cabinets and reports that are unequalled anywhere; remembering Henry Ford who insisted that "history is the bunk" even while he hired college kids to do his homework for him; embarrassed by the memory of Henry Ford thanks to an historical scholarship today unmatched in excellence; possessing finally a sophistication that is truly cosmopolitan, America is still that aging athlete who knows all about sex and who is nonetheless impotent. America cannot perform within history. America—the American Northeast, not the defeated South—lacks any corporate historical imagination, any capacity to live within a time larger than itself but one with western history. This parochialism is today dying in our streets and in Vietnam. Our search for our own Past, whether it be Paradise Lost or Paradise Gained, is America's Future. Footfalls may echo in the memory down a passage which we did not take and towards a door we never opened. But there are other footfalls which open into the rose-garden that we did take. And if the roses be withered today as the century approaches its winter of desolation, it might very well be because the future is contained in time past.

The Symbol

Change It at Your Own Risk

The American Cowboy

And *Law and Order*

ONE of the better kept secrets of our day is that Dr. Marshall McLuhan, the high priest of the new communications revolution, is a philosopher. This is not surprising given his long association at Toronto with the most illustrious philosophers this century has known. This has made him a philosopher of a certain kind and of a certain school. But McLuhan hides his intellectual affiliations so well that we have reason to suspect that he does not care to spell them out. This possibly would be too "hot" and therefore we shall not "spell out" — this would be print technology and hence "high definition"—what is there for anybody with philosophical sophistication to see.

Nonetheless the theme of his book, *The Medium Is the Massage,* has become a slogan in both periodical and popular press. McLuhan is paid the high compliment of having his slogans used by men who do not cite him and who, in some cases, never even heard of him. Better tribalization could not happen to a man who insists that the sharp identification of authorship belongs to a book culture, dominated by type, which is dying along with machine technology and the rationalist mind that gave it birth.

We begin our meditation, then, upon the meaning and use of symbolism in contemporary and modern western society by noting that McLuhan's "the medium is the message" felicitously expresses the distinction between signs and symbols. This last is crucial for every student of culture. Signs—much as red and green lights and red-and-white candy-colored barber poles—do not participate in the reality they signify. Content with being pure "pointers," they are grasped by the sensibility, taken in by the mind, and then immediately transcended in the mind's coming to grips with the signified.

As José Luis Aranguren puts it in his *La Communicación Humana*, every communication involves four elements: (a) the sender, (b) the transmission belt, (c) the receptor, (d) the message. In a pure "pointer" or sign situation, the message is transmitted to the receiver by way of a "belt," be that belt language, gesture, or any other medium of communication. In a *pure* sign-signified situation any old transmission belt will do. What is important is the message. If a ship is sinking, Morse code, radio, semaphore, flags, or distress flares will equally, circumstances being the same, get across the message from the sender to the receiver. In the pure sign-signified situation the medium disappears before the message. However, the "transmission belt" itself or the medium is a state of reality which sets up a *mode* of being, englobing sender, receiver, and message. To the degree to which the medium imposes itself in culture and increasingly identifies itself with the message, the medium becomes symbol rather than mere sign. This is most forcefully illustrated, within communication theory, by the sacramental sign which is actually a symbol as we have defined the term here.

The sacramental sign, in traditional theology, actively effects what it signifies. To a believing Christian, the prescribed words spoken plus the pouring of water *causes* baptism. These signs are not incidental (in the theology in question) nor could they be altered at will by any other signs. Unlike smoke as signifying fire which is something seen and which could be substituted by either smell or touch, both of which also carry the message of fire, sacramental signs are not indifferent to the reality they signify. The grace of baptism, achieved by water, is dependent on the use of the proper sacramental signs. There is no divorce possible between *what* is said and done and its "being-said-and-done": being is one with saying and doing. The sacramental sign causes what it portends, because of the power of God. This last points up the relation and difference between sacramental and magical signs. The latter bring about what they promise but they do so by their own power. A voodoo doll stuck full of pins kills the offending person represented by the doll. A rain dance by Navaho Indians, through its incantation, calls down rain upon parched soil. But every true symbol partakes of the magical. Symbols blend into what they signify, participating, as Paul Tillich insisted, in the reality signified. Let us distinguish, therefore, pure signs from symbols by suggesting that whereas a mere sign which functions as a pointer can be altered at will without changing its signification, any alteration in the symbol involves a change in the signified. *Lex orandi est lex credendi:* the medium is the message: *Bild ist Bedeutung.*

Something magical enters every cultural symbol by which a society expresses its encounter with reality and incarnates its way of life. These symbols which not only cluster around a society and gesture its secret to the world are altered at the peril of that community's very corporate existence. Now media are symbols of this last kind. Media remain "transmission belts": i.e., they relate messages of all kinds; books tell all kinds of stories; language is used for an infinite variety of purposes. Nonetheless, media, by establishing themselves as modes of existence within which cultures live and move and have their being, invade the psyche and shape the sensibility and nervous system according to their own patterns. Working their magic into the corporate consciousness they alter the psychic response of man to reality.

The medium is a pure transmission belt or *sign* only when it is rigidly and artificially controlled as in the spoken and written expressions of scientific formulae used by physicists. Abstract concentration in such instances reaches such a peak of intensity that the mind totally identifies itself with the *other* —meaning with the meant—that whether the formula be written in chalk on a blackboard or relayed through a deciphered code is irrelevant. Let it be noted, however, that in life, where abstracted attention is diffused through a nervous system bathed in the whole gamut of sensation and emotion, media create their own meaning or message which is one with their being there at all. Signs, thus, take on a symbolic value which is violated only at the risk of betraying the initial meaning given them as mere signs. The best instance of this is profanation. A flag, initially nothing other than a "pointer-sign" indicating to an army the location of its chief on a battlefield, came in time to *symbolize* first the chieftain-king, then the kingdom, and finally the nation. Pius IX reproached the Count de Chambord for refusing to accept the French throne in 1870 because the Pretender insisted upon a restoration of the Fleur-de-Lis in place of the tricolor of the Revolution. What to the Pope was "a rag" to Chambord was a sacred symbol: to him the medium was the message. All media, therefore, partake of magic because man's involvement in them alters him willy-nilly. This alteration affects the very messages he transmits because he is not the same person he would have been, nor is his reaction to life what it would have been had history not thrust him into *this* world, dominated by *this* set of media. Nowhere is this more vividly manifested than in language. Words are so charged with symbolic significance that very often they seem to almost live a

51

life of their own. The effort to empty language of its historical cargo achieves occasionally a one-to-one communication between transmitter and receiver, but it does so only by laying language on a procrustean bed formed of abstractions. It is a commonplace that all languages reflect national psychologies and any totally bi-lingual man is a schizophrenic in that he possesses two personalities rather than one. The same message fed through two distinct linguistic transmission belts is going to come out altered. "Sign-pointers" are inexorably the victims of symbolic media. This is so true today, because of television, that an event unreported becomes a "non-event" and "non-events" can be called into existence at will by the masters of the tube. "Being" is magically obliterated and what "seems" but is not now comes "to be." Thus the televised symbol effects what it signifies and electronic symbol shades into a new magic. Even revolutions are made and unmade within the studio by sorcerers' apprentices bent on witching into being a new creation.

52

The power of the symbol over the signified today has become almost absolute. We are living, in the words of Pope Paul VI, "in the age of the image." Every tribalized community must exorcize that which lies beyond it in the name of its own image. He thus remarked the essence of tribalized life. Exorcism bewitches out of being whatever, if permitted *to be* , offends the tribe's conviction that it incarnates the fullness of truth and even being. In an analogous fashion the family chases out the stranger as a menace to its own internal cosmic meaning. Lovers flee not only the presence but the very being of anyone who would intrude upon their love. Now electronic media both create, and are consubstantial with, modes of being proper to themselves. It follows, therefore, that whatever is admitted to the magic electric circle *exists;* what is not admitted does *not* exist. Let it be noted that we are speaking here in an absolutely formal and rigorous fashion: if a medium is a form of being, then anything lying outside that medium is non-being *to the medium itself* . In order to see that any given medium is only a slice of reality a man must be above that medium and must synthesize it into a whole involving all media. Media cannot synthesize themselves. A medium is as totalitarian as is a cliff—beyond it the Abyss, the Nothing. As long as the medium remained the magic circle of friends gathered under candlelight in an inn, united in comradeship against the world beyond; the village tucked within a valley shutting out whatever might lie outside; or the tribe sharply confined in space, the absolutizing character of all media was checked by a mankind using them all but absorbed by none of them because humanity itself was identified with no one of them. The solipsism implicit in all media was controlled —by Man.

But electronic media have annihilated space and are rapidly abolishing time. Plato wrote that the Polis was Man writ large. Today electronic media are self-enclosed worlds writ large. What happens outside them is as unreal as that which happens beyond the embrace of lovers to the lovers themselves. It simply is not. Were it to be, the medium's claim to being would be abrogated. Television signals into existence that which is not and annihilates—renders it never to have been— that which is. We enter today into a weird world of fun-house mirrors in which *being* is not and *seeming* is. Given that the new media surround the entire world even as it creates a new one, we are forced to ask ourselves the truly dizzying question: what is real anyhow? Ask any actor: he will tell you that the play is the thing!

The new media attract to their service as would a vortex the kind of men for whom reality "as it is" does not satisfy. Thus they make real what was only a dream for Mallarme. Their wands call into existence worlds that never were, and the waving of their wands bewitch into nothingness the world from which they fled and which once was. This power reaches even to the outer recesses of the human spirit. It can even destroy a world religion, the Catholic, today being systematically ruined by media within which that religion exists as it never was and within which what it once was has ceased to be. To bathe oneself totally within any medium—be it radio or press or television—is to systematically obliterate the possibility of transcending the medium in order to integrate it into a larger vision. Signs transmitting messages tend inexorably to convert themselves into symbols participating in messages. Very soon they become the message. The great misunderstanding today about media is one with a confusion between messages sent and transmission belts. An attack on the transmission belt is usually an attack upon what is transmitted; an attack on signification is generally an attack upon the sign: television is bad because its offerings are vulgar. Therefore, let us shut off the "TV." Nonetheless the same people launching such a criticism would not condemn the Bible because it is something written; nor would they condemn music because of the Beatles. Moralism will not help us here. Insight into media is one with knowing that all media set up states of existence that tend inexorably to deny existence to any order of being lying outside them. Signs signify but signs try to *be* because even signification strains to exist. The pagan god of Aristotle only thought: he was pure signification; but a new God, Christ, said that He was: "Before Abra-

53

ham was, I am." The medium *is* the message and the gradual perfection of media renders this identity all the more perfect.

But the older rationalist mind which even today survives and mingles with the more contemporary consciousness of an age being shaped by electronic media denies the above contentions because the older rationalism understood the symbolic order in a different way. The rationalist method, elaborated by Rene Descartes, was suffused with the conviction that mathematical knowledge is the archetype of all knowledge. Given that mathematical concepts—as then understood— participated in a kind of purity and clarity of abstractness denied other articulation, Descartes insisted that the mind begins by intuiting pure ideas after having cleared itself of the garbage of the imagination. Given that sense qualities cannot be grasped in pure ideas—who ever had a clear and distinct idea of red or soft?—sense qualities were suppressed in favor of their symbolization in mathematical constructs. Given that the symbolic approaches the density of existence and the richness of the sensible order, it had to be purified until it became little more than a system of pure pointer-signs. The laws of mechanics cannot be expressed in a dense and mysterious symbolic context. Such laws—and rationalism saw the entire cosmos as being constituted by them—must be pointed at with the most spare and lean language. This jargon is the more perfect in proportion to its ability to disappear before the mechanical order. But if symbols are mere sign-pointers, then it follows that rhetoric and grammar are no longer contexts *within* which man grasps meaning and which orchestrate it by an integral participation. Grammar and rhetoric were forced to become the outer clothing—ideas put on in order to appeal to the emotions. This was a sop thrown to a body already served a summons by mind to appear in the divorce court. The symbolic order was reduced to *exhorting* man to action. Earlier the symbolic order had been an indispensable cause, not only in exhorting man to do, but in *teaching* him to know. The profound difference not only marks two distinct approaches to education but it also suggests two irreconcilably opposed stances before the real. Rationalist thinking confused men even as they were being shaped by new media produced by machine technology and linear thinking. Rationalism absent-mindedly created a new world without having at hand any intellectual instruments enabling it to come to grips with what was happening to the West. Ideas—understood as repeatable mechanical models—were divorced from the symbolic. Men thought they could symbolize any old meaning by way of any medium at hand without that medium affecting profoundly the meaning itself. This, of course, is perfectly logical if there be no distinction between signs and symbols, being pointers at, and participators in, reality.

Aristotle knew this: science uses the particular to reach the universal. But the arts grasp the universal—meaning— *in* particulars. Dramatic action is just *that* —action. All action is concrete and existential. Literary value and meaning is always seized in particular actions. Genres were classifications of these actions. But the Aristotelian insistence on finding literary intelligibility in particular actions was lost to the criticism which grew up under the aegis of rationalist thinking. Conceiving genres as though they were logical or formal patterns laid up in heaven, the humanists attempted to impose them upon media largely unaware that media not only alters the message but in a profound sense *is* the message. A sonnet read and a sonnet sung are not the same sonnet. Shakespeare played on the Elizabethan stage and Shakespeare read out of heavily-footnoted books by nineteenth- and twentieth-century students are not the same Shakespeare. The action is not the same and if we are urged today to go where the action is we are thereby being urged to go where meaning is, where the media are. This attempt to treat media as instrumental signs rather than organic symbols of meaning can be illustrated by the critical confusion surrounding the motion picture "The Graduate." Critics who approached the picture armed with the rigid categories of literary genres found a perfect comedy: characters do not develop significantly; the ending is happy; virtue triumphs thanks to the emergence of "The Girl"; etc. But audiences unburdened by literary pretensions saw in the movie simply a vivid presentation of bed-romping and went home, alternately delighted or disgusted, depending on whatever moral and aesthetic sensitivities they brought into the theater. The critical or literate reaction was formally correct but existentially irrelevant; correct formally because the comic genre *is* present in the film; existentially irrelevant because this genre had as much effect on the total response of the audience as does a piece of tape glued to a brick wall upon a person contemplating said wall. The very perfection of the medium itself simply swamped the genre and persons attuned to media rather than to literary forms, some ninety percent of the people who saw "The Graduate," lost any comic genre contained in the movie. Again, "Bonnie and Clyde" was blood and guts to a media-tuned generation. Any "heroic" structure intended by the writers of the script was dissolved in the technical perfection achieved by the motion picture industry today. The "blood and guts" were very bloody and gutsy

55

indeed: they had nothing to do with the heroic sense given the phrase by an earlier generation. Rationalist criticism invades the existent with structures treated as though they were Cartesian beds upon which media could be stretched. But media will not be so stretched because they are not mere signs that point. They are symbols that participate in meaning and that therefore *are* meaning.

T.E. Hulme threw a bomb into literary criticism when he suggested, sometime before World War I, that the language of science—being abstract—is sloppy whereas the language of poetry—being concrete —is accurate. His young woman walking down the street with her high heels click-clacking and her skirt swishing back and forth with the movement of hips signalled the return of symbolic meaning to existential reality.

Hulme saw that concrete meaning is always grasped in beings. We might add that meaning is grasped in things even if they be media. It follows therefore that a double motion in extremes is permitted the human spirit and sensibility. The thing can become a pure symbol of meaning. Meaning can be lost totally in the thing. Either Hulme's young lady is a pure sex symbol, or she is nothing other than a gal going down the road. When this last occurs man loses himself in the stream of becoming. He is submerged in a pure Wagnerian romanticism in which the waves of being lap over his head as he plunges into combers promising him nothing more than identification with mother ocean lapping him into her bosom. When the former occurs, Hulme's corseted young lady stiffens into a puppet, all cardboard, that bobs back and forth signalling only the Lord knows what deep and dark meanings hidden from the prescience of ordinary flesh. Hulme's insistence that real poetry "ticks off" exactly that which is, reiterated the Thomistic teaching that the modes of knowing follow the modes of being.

Most contemporary criticism has not followed Hulme's lead. Instead of finding meaning in being, in artistic action, it has been tempted to look upon artistic action as a simple vehicle for meaning. This was most particularly illustrated in the Marxist "social criticism" of the thirties. The entire aesthetic order was subordinated to political and sociological goals which drained art of its irreducible concreteness. Even though this sociological aesthetics, due to its very crudity, no longer commands the attention of the academy, its residue still clutters departments of literature everywhere in our nation. Professors of English, moved slyly by the puritan mystique which insists that all activity be pointed towards salvation, justify their professional existence by dissolving the marriage between literary meaning and being;

they seek salvation from the written word; they make literature do for philosophy and theology. As Cleanth Brooks has stated privately, novels become substitutes for religion and the graduate schools of literature around the nation are glutted with young folk seeking salvation from reading poetry. The aesthetic does duty for human and divine love. Contemporary criticism—even as it has fought valiantly to clear its head of the Cartesian hangover—is tempted to see works of art simply as clusters of symbols. The decline of the book in our time has conspired in this transformation of goals. Who today reads a great novel for sheer enjoyment? Who does not hope to find some clue to grace in poetry? Students in university and high school are urged to hunt for "meaning" until the very word has become repugnant to millions of Americans who say the hell with meaning! Give us reality! Huckleberry Finn did float down the Mississippi and Mark Twain fulminated in a famous paragraph against anybody finding anything more or less than Huck's floating down that muddy river. This has not prevented a score of doctoral dissertations from being written on the meaning of *Huckleberry Finn*. Robert Louis Stevenson is out of favor nowadays because he was such a good story-teller that the reader has little time or inclination to look for complicated symbolism. The symbolism *is* there, no doubt about it! But it is consubstantial with the tale told! The rationalist mentality linked with the Calvinist insistence that we squeeze justification out of the written word forgets that things and actions, while they adumbrate transcendent meaning, also bend back and finger those very things in action. Scott's Ivanhoe symbolizes courage but courage is symbolized in Ivanhoe. Faulkner's *The Bear* points beyond itself to meanings hidden to all but the elect, but anyone from Faulkner's part of the country knows that the Bear was indeed just that—a bear.

Normally the mind moves easily back and forth between symbol and symbolized. Children constantly confuse both and thus confess that they are normal. Mother is warm; therefore warmness is mother. Security is to be held—to be held is to be secure. This interplay simultaneously intensifies both the density of existence and the complexity of meaning found therein. Intelligibility (by now the reader must be tired of the word "meaning") is encountered within concrete actions and things, persons and places, that are one with the irreducible action that is being. The most compact instance of this in the juridical order is the Hungarian Crown. In Hungarian legal theory the Crown not only signifies the nation but quite literally *is* the nation: the lands of the Holy Crown of St. Stephen belong to the Crown, not to its wearer. If the medium, in this instance, were not totally the message, the United States would not have taken such pains to smuggle the Holy Crown out of Hungary and away from communist masters who hoped to legitimize their rule before the peasants by displaying the Crown: Hungary.

Political and social consequences of the shift in the role of symbol is illustrated by the fate of western philosophy. The sharp distinction of man into "subject" and "object," into what we shall call role bearer and role, is most vividly seen in Leibniz for whom man is a self-enclosed monad determined by an indefinite number of predicates that happen to him because a *deus ex machina* so wills. Heavily influenced by printing technology as well as by the isolation within which philosophers worked in the seventeenth century, the distinction of man into a "subject" that receives "objects" was at least partially a consequence of the spatial division of written sentences into subjects and predicates: the subject, after all, is grammatically distinct from the object.

If man be principally a subject that picks up grammatical objects, it follows that this symbolic structuring of the psyche illuminates the French Revolution and the Age of Liberalism that it fathered into being. The revolutionary Rights of Man are objective and abstract characteristics that properly ought to happen to every "subject" understood now, not as a man subject to authority, but as a substratum that acquires rights. The shift from the older classical political theory could not be more distinctly emphasized. Classical political theory saw man principally as owing duties to his fellows. The newer order viewed him as a spatial center of rights coming to him from the peripheries—from the right hand side of the folio. These Rights were discovered by encyclopaedists who did their homework literally "at home" under candlelight. They were seated before a desk with a manuscript text in front of their eyes. Rights were not rights until they were written down. Hence the insistence from 1789 until today that all political constitutions be *written* constitutions. The joke about the American asking the Englishman for a copy of his constitution is too old to stand lengthy commentary, but it does point out the deep American conviction that the political order is principally something found on a piece of paper and then digested into a cerebrum. The dozens, if not hundreds, of South American and European revolutions during the past one hundred and fifty years also grew out of a belief that political life can be manipulated as easily as words and sentences can be altered in written composition as they are juggled about within a spatial imagination. Hence the fervent cry of Spanish Liberals in the early nineteenth century: *"La constitución o la muerte"* —The Constitution (i.e., the written constitution) or death!

The worship of the written word as a political symbol is most especially true of the American experience. Although we must never forget the survival of older medieval elements (e.g., the sheriff and bailiff, common law, municipal self-government, etc.), America—in quite a literal and literary sense of the term—was founded as a nation by being written down on paper. From the Mayflower Pact, through to the Virginia Declarations, down to the Declaration of Independence and the Constitution, America hammered out its corporate meaning in print. As if this were not enough, "Publius"—that hydra-headed author—felt it necessary to issue a document which interpreted the Constitution: *The Federalist Papers.* More ink has been spilt over the meaning of *The Federalist* than has been spent on elucidating the mysteries of the Book of Revelation. Students of American politics will refer Chapter V of the *Federalist* forward to Chapter X and Chapter X back to Chapter III in a valiant effort to come to grips with what America means or at least meant in the past. A tradition that is so heavily cerebral must be constantly debated. But when meaning is simply a condition for being, for existence, the situation alters radically. Nobody debates about who he is. He accepts the fact just as dwarfs in other cultures accept their fate, their meaning, as the condition of their life. And everybody else takes them for being just who they are and as they are: dwarfs. "Being a dwarf" is not a state that happens to one. It is just part of his existential condition and he knows this as do his fellows. This makes the dwarf's real tragedy easier to bear. So too with cultures: those that live out their meaning without articulating it in print never have any doubts about what they are within the economy of history. Meaning, for such societies, is simply a function of being, a style of existence. Meaning has no separate reality. But the politics of fragmented and abstracted thought are the politics of objectivity "happening" to subjectivity, meaning accruing to persons. These objects are capable of being manipulated and reinterpreted as are all written objects. It follows that America has been one long and often-brilliant debate about her own meaning, not only as expressed in, but as being one with, the Constitution, our principal political symbol. There are parallels in Christianity's constant debate about the meaning of the Bible but given the dominant secularist tone of the American experience these religious debates never erupted into civil war. Our Civil War, the bloodiest in history, was fought over the interpretation of the written text of the Constitution: our equivalent to the Nicene Creed. Until very recently the one untouchable idol in the marketplace was the Constitution. No drunk dared insult it in the lowest bar in the nation. Everybody had to be for the Constitution no matter how bizarre an interpretation he gave it. At no time in the entire history of mankind has a written political document been elevated to such a sacred status. The American message was the medium.

De Tocqueville noted with his customary shrewdness that the written word accompanied the American frontiersman everywhere. The Tradition was enshrined between precious covers for families that lived on the edge of violence and within the shadow of the forest: the Constitution, the Bible, Blackstone—these followed the frontier family into the very zone of perpetual and daily danger. Our human symbols from Abraham Lincoln to Horatio Alger have been of poor boys struggling to get an education, to learn how to read and write. Lincoln did not learn to read by writing on a shovel with a piece of coal but the myth here is closer to truth than is reality itself. Our heroes were anonymous subjects acquiring predicates; along with them they gained acceptance and value: the tougher the road, the greater the premium.

Tradition in the public American experience is less an act of being lived, a subjective density, than an objective content discovered by searching through documents which are then put to work in the community. Following the pattern of abstract thought, there emerged a sharp distinction between the symbolized and the symbol. Our heroes have usually been revered for what they *do* rather than for *who* they are. In less-thoroughly mechanized societies the sign and the signified blur. They become a genuine symbol as we defined the term earlier. The king is revered simply because he *is* the king. His legitimacy accrues to him for being the son of his father, not for any personal merits that might belong to him. Americans find this last curious if not monstrous; yet bloody wars have been fought by highly-civilized nations precisely over the principle of monarchical legitimacy. The issue has nothing to do with the value we attach to it morally; the issue has to do with understanding differing ontological experiences. Parents today, however, are revered for being parents rather than for acting out a "parental" role. But wherever the symbol is strongly differentiated or fragmented from the symbolized, the carrier of symbolic meaning is aware of a distinction between who he is and his role. Trying to "live up to that role," he works very hard *to be* a father or a mother or an executive. Vaguely guilty because he knows in the depths that he is not what he would be, man experiences a fissure between his being as given him in existence and the meaning or role he incarnates. Professionalism always accompanies this schism between being and meaning. The professional mother, that very familiar American phenomenon, works so hard at being a mother that she produces the Momism made famous by Philip Wylie many years ago. This agony of sensing and suffering that we are not what we pretend to be reaches its apotheosis in the office of the American presidency. He usually breaks his heart and his health in trying to be what he is not by nature, in attempting to convert the objectivity of the meaning of the presidency into a subjective density exercised by a person who simply is not president by any existential legitimacy of being. The so-called "identity crisis" in the United States is little more than an acute awareness of this hiatus between being and meaning, personal subjectivity and objective responsibility, existence and essence. The traditional European scholar almost cynically contradicts this American attitude. He is a man who is so identified with his profession that he often feels absolved from working hard at it. His status possessed permanently— the antithesis to the "job" which is an attribute taken on but never really possessed, owned—so blended meaning and being that it achieved legal sanction in many European countries: the holder of a chair cannot be dismissed in Spain even if he offends General Franco. The very concept of "being fired" is foreign to this archaic compacting of sign and signified, of role and role-carrier. Signing a contract or winning an election within a society based upon rationalist contract rather than organic status involves taking on a role, assuming a list of responsibilities considered to be crucially important publicly and objectively but foreign to the private and psychic life of the subject who assumes them. Therefore they can be cancelled. To be fired or voted out of office is a logical consequence of the mechanical fragmentation of subject from object.

61

The Latin tradition has been schizophrenic in this regard because rationalism, itself a Latin invention, was grafted heavily onto an older organic order. Government was centralized under Louis XIV and laws were dictated in minute details that imitated the mechanical laws supposedly governing the universe. But the Latin legal machinery written down in statutes has always been subverted easily in life. The law is severe but its severity is applied only by fits and starts and in moments of crisis or bad temper. Nepotism is normalcy. Any self-respecting Italian or Spaniard would think it a sin not to get a job in the firm for his brother or cousin. Familial relationships are deeper and older than the newer abstract and revolutionary superstructures. These last must be circumvented in the name of the tribe. The government, after all, is "over there"—our enemy! Americans often do the same thing but they are ashamed of doing it. Fair is foul and foul is fair, depending on how deeply mechanical thinking has taken root in a community.

The difference between being purged of meaning in one's life but bent on acquiring it by sheer will power is illustrated by the subtle relationships that govern intercourse between rich and less rich in the United States. The children of the rich often pretend that they are not rich when mingling with their fellow students at school. Yet everybody who has inherited money is happy about it because everybody is happy to have money, which, as Hilaire Belloc once wrote in a delightful little ditty, is the only thing that "pleases all the time." But the American rich boy (euphemistically called "wealthy" by the middle classes) is guilty about having money because he did not earn it. He did not bridge the gap between subject and object. Now inheritance is a cherished Anglo-Saxon legal ideal but until very recently (we shall return to the theme later) inherited wealth has set a man aside from the dominant ethic which placed a premium on going it alone and making it on your own. This is why rich fathers make their boys work their way through school. This is why the guilt feelings of the rich, accentuated by omnipresent media, have forced them into a fierce search for privacy and into driving middle-class automobiles. Even the Rolls-Royce democratizes itself today in a bow towards democratic egalitarianism. Americans—rich and poor alike—despite political and ideological commitments, find every kind of government intelligible from Communism to Establishmentarianism, even when distasteful. There is but one exception: hereditary monarchy. The role of king is inherited and the man exercising that role does so because of *who he is*, the son of his father. His meaning in life flows out of his being. But meaning that flows out of being rather than being that flows out of meaning is highly un-American. Anything else, in the heyday of the American dream of rags to riches, a dream which hangs on as a national symbol even as it loses reality in the market place, was something less than moral.

This attitude has helped produce the great crisis within the American family. The anxiety quotient of the American woman is rooted in her conviction that she must win what she already is: wifehood and motherhood. Victimized by the need to play roles, the American woman in upper suburbia must simultaneously act out the part of mother, wife, lover, educator, civic leader, fashion plate, chef, hostess, interior decorator, glamour girl, scrub woman, and garbage collector. All things to all women and to all children and at least to one man, she senses dumbly as she pedals a perpetual treadmill that she is nothing to herself. She cannot relax despite all the advice shouted at her from magazines and television by white-coated custodians of bodily and mental health. To relax involves sinking into something but she cannot sink into the nothingness covered by her dozen facades. She has juggled masks so long and done such a superb job of it that she has become her roles. She has objectified herself out of subjective and personal existence. Eliot ought to have written another poem, one about "The Hollow Woman." It would have had a line on meaning passing for being and being vacated of itself in a supreme sacrifice offered to a mystique of achievement that permits nobody to mean simply because he is. Possibly Popeye is a protest against all this role posturing: "I am what I am and that's all I am—I'm Popeye the Sailor Man."

The need to win significance, to achieve the "status symbol," points out the rationalist marshalling of events in series which fall under some common type. When meaning is thought to be a function of being or existence it is impossible to line up realities under a common type. There is no blanket which we can throw over a cold glass of beer, a missionary in Africa, a combo band, and an automatic rifle. These things are absolutely different and their only likeness consists in an analogical and relational in-depth intensity within themselves to their own in-built meaning, to their styles of being. This kind of analogical knowing is foreign to standardized thinking in which abstracted essences or "meanings" become the yardsticks by which individual "instances" are measured and judged. Analogical diversity gives way to a mechanical line-up well illustrated by suspects in a police station or soldiers on a drill field. The need of the person in our culture to measure up constantly, to shape up, to put on the mask, produces a gnawing anxiety without surcease. The distance between reality and ideal, between personal subjectivity and imperative objectivity, forces men constantly to reach, to outdo themselves. Am I playing the role right? Am I pulling it off? Am I a success at what I am doing? Is my face put on straight? Did I start from nothing or at least pretend to start from nothing (am I a poor rich girl making it on her own?) Do I, says Willy Loman, live up to the territory? The strain under which such a civilization lives takes an enormous toll in wear and tear; it must then keep up a tranquilizer business. It is also the price paid hitherto for a society geared to high efficiency and production.

The convictions clustering around any society are usually first played out symbolically before they are articulated rationally and thus known to be symbols. Historically symbol is reality acting before it is known formally as symbol or meaning. We need only recall Vico's contention that the social realities of one age become the ideals or symbols of the succeeding: thus chivalry had ceased being a reality in the rest of Europe when it passed to Spain; thus the times of Homer were symbolic ideals for the Greeks of the Golden Age of Pericles. Huizinga insisted in his *Homo Ludens* that meaning is lived and played as exercised act before it is articulated. In this context the most significant American symbol has been that of the Frontier. America's destiny, involving the conquering of a hostile frontier, was forced on our ancestors by the exigencies of nature. The early Americans moved west and hacked their way through the forests or they stayed on the beach and prayed for the next boat home. There were few alternatives. The Frontier, a physical reality to those early settlers in the New World, soon blended with the American need to conquer and to win meaning on one's own. The rejection of European hierarchy and order, built on status, was imperiously forced upon the colonists.

Take away from an American his promised Frontier, his need to gain or win, to conquer, and that American is in trouble. Americans retired on pensions too soon tend to go to seed as expatriates in South American countries and elsewhere. They cannot stand all the leisure. It demoralizes them. Similarly American tourists, bent upon "doing" Europe or "seeing" South America, must get through a certain itinerary and conquer a Frontier, visualized as a field of battle to be won. The classic American drive to conquer the Frontier was spatialized westward because the colonists had come west from out of an east, Europe, that would forever remain for them what lay behind their backs. This movement west took on such a moral force that even within the confines of the United States the westerner was looked upon as being more virile and noble than the easterner whose occasional excursions into the west made him an object of humor and ridicule. The dude rides horses badly and his fastidious clothes are effete. He can't shoot straight. The man from Boston looks bad in Bonanza Land.

Kennedy's New Frontier was the elevation into political myth of a national hunger that had already been sated as reality. America first ran into trouble when it reached the Pacific beaches of California. America had no place to go. And unless an American can "go" he is frustrated. And when "Go west, young man!" ran out of objective correlatives America was forced to search for new ones for its drive. Did poverty not exist in the world to be assuaged and were the moon not there, we would have to invent them both. We cannot stand still. The Frontier differs profoundly from the European understanding of the term which is taken from walls such as Hadrian's in the north of England: it kept out the Picts. The European wall—we need only think of the perfectly walled city of Avila on the high Castilian plain—"keeps out" just as English and French picket fences do just that: "fence out." The European crosses the frontier only in isolated moments of adventure or in order the better to protect what lies within. The entire Roman expansion into the deserts of the Middle East and the forests of the Germanies in the last centuries of the Empire was either the result of some obscure call to glory in a few soldier emperors or, what was more common and steady, the consequence of a need to protect the border provinces of the Empire from incursions by the barbarian. Rome could retrench and give up frontier borders when immanent danger passed. There was no loss of prestige in doing so. The periphery existed for the sake of the center. But the American Frontier is something altogether different: it must be crossed because salvation lies upon the farther horizon. Being is always somewhere else, never where one is.

65

Today the Frontier as a national myth is challenged by, and is dying before, a new cluster of symbols centering around peace and prosperity. The American drive is dispersed through goals such as foreign aid to underdeveloped countries and the space program. The disappearance of linear or horizontal frontiers gives way to those which are vertical and to those which cannot be spatialized at all. But these symbols as yet are shifting and the progressive decline of the old work ethic under our metamorphosis from a producing to a consuming economy has left America deeply troubled about her own meaning. Unwilling and even incapable of articulating its destiny in terms of the charismatic, America today wants a frontier but can find for it no adequate content. The skeletal frame of a symbol hangs on the national wall but no picture fills it.

Where the abstract mind inherited by the West from the Renaissance dominates and where the machine technology produced by that mind imposes itself imperiously, corporate and social symbols are fragmented and are dispersed into instances of types rather than coalescing into concrete realities. American status symbols, for instance, hardly ever are *things* charged with significance. Usually they are instances of species or essences: not *this* Cadillac, but the Cadillac as a type and as a banner; not *this* house but this kind of house; not *this* neighborhood but the class to which this neighborhood belongs. The European cleaving to things—to ancestral houses, to a particular bit of soil, to one car loved and cherished with pagan affection through fifteen years of use—bespeaks an engagement in the material world that has never interested a nation that has always been on the go, that has always been leaving its past behind itself as rapidly as it turned in last year's model for next year's. The first thing an American couple needs to know about a house is its resale value; the first thing a college graduate wants to know about a profession is its retirement benefits: he does not sell out at forty; he signs up at twenty-two with an eye towards those golden sixties. Given to transcending things themselves as part of the national style of life, Americans do not dig into matter. We cannot stand dirty fingernails. The national mania for cleanliness bespeaks a fastidiousness the origins of which mingle with an older puritanism and a forever dominant rationalism. Eric von Kuehnelt-Leddihn has suggested that Amerians are the most conservative and least revolutionary people in the world. We want more and more and we want the bigger and the better of the same things. Novelty we abhor. As a nation America has absent-mindedly produced an enormous technological revolution, but America does not today possess the sensibility necessary to understand what it has done. We do not want what is absolutely different. It follows that our style of life is univocal and not analogical. Because univocal, our humor tends to be of a certain kind: we delight in men and things that fail to live up to their types. American humor does not concentrate upon the idiosyncratic as does the English but upon a failure of measurement. This falling away of things from their types is the countermovement of the work ethic which insists on reaching up to types. American humor is a comic relief from the mystique of production and success. Full of slapstick and very good fun indeed, American humor lacks ironic and cruel overtones. These last better fit an analogical vision of the real. Tacitus' crack about a Roman emperor, *"Capax imperii nisi imperassit"* —he would have been an excellent emperor had he not been one!—is quite foreign to the American mind.

This cerebral character of mechanical and highly-literate thinking hides a certain irony. Bewitched by the *idea* of matter which is somehow mingled with the myth of pragmatic efficiency, most Americans—at least after they have sown their wild oats—do not want to be engaged directly in matter. The high quality but general tastelessness

66

of our processed and chemically treated food as well as the national horror of B.O. indicate the suppression of a balanced sense ratio in favor of the totalitarianism of sight. Sight is the sense which is closest to the cerebral and, according to some thinkers, the sense which is closest to the soul. The American over-emphasis on sight is most clearly evidenced in sex. The American man can look all he wants and he is even bombarded by the media to do so. Playboy clubs have made a business out of looking—but don't touch that Bunny! And Bunny! don't give out your phone number! Tail pinching, an honorable part of Italian street life, is frowned upon by Americans who make a national habit of what old-fashioned theologians would have called "concupiscence of the eye," as they cerebrally feast them upon strolling feminine pulchritude on summer beaches. Look, but touch only at your peril, and—for God's sake!—don't smell! Television in turn will tell you girls how to cover up *your* smell which is indecently personal and will instruct you in a smell which fits your type. It is interesting that the perfume business in the U.S. owes so much to France, the nation that gave birth to Descartes and the mathematical method. Smell excites primitive peoples and even less-rationalistic occidentals, but no American in his right mind would want to go to bed with a hot and sweaty girl just off the tennis court.

67

Rationalist cerebration mingles, however, with other symbols whose roots are more organic and even quasi-mystical. These last are linked with half-forgotten and older dreams of rootedness. They are often as muted as is the memory of a landed gentry that vaguely haunts the suburbanite with his lawnmower and his own grass to cut every Saturday. But even here rootedness gives way to the vastness of the frontier. The barbecue pit, unique to the U.S., remembers the chuck wagon days of Bonanza Land. The cowboy, today under heavy attack, still remains *the* great National Hero for fifty to a hundred million television-rooted Americans. He is the American knight, our man on horseback. A rigid set of aesthetic and moral principles governs the pure cowboy classic. Changed by innovators, the type remains nonetheless clearly discernible. The cowboy rides into town from nowhere. He has no family and is totally without ties. He finds the forces of evil and sets right everything wrong in town. He puts down the cattle baron who exploits the people or he takes on the gunslinger and calls him out at high noon. The cowboy is always single. In the pure classic form, although he could get the girl at the end, he rides away into the horizon as untouched by responsibility and by familial rootedness as when he first tethered his horse before the saloon door.

Just as the Arthurian legends underwent transfigurations and still remained themselves so too has the American cowboy. Even when he settles down he retains his mythological origins. In Bonanza, Papa is a widower and his two eligible boys are unmarried. The Rifleman also is a widower with one son. Matt Dillon, Marshal, is loved by Kitty but resists marriage. Does this not suggest the American male's secret? Does he not wish, without knowing, to wash himself free of the female? The chastity of the cowboy guarantees his liberty and keeps open that Frontier over whose horizon he must ride tomorrow.

But the cowboy's aloofness from women is something negative that assures him his way of life. His purity is not religious. It does not grow out of a vow and it does not suggest that deep medieval conviction that chastity is a positive quality making a man more manly. The cowboy's aloofness from women is strictly functional to his self-imposed loneliness. The classic cowboy is shy and timid. When he gains the girl in the last act the genre is offended. This timidity reflects two truths about the American man: the cowboy is attracted to women and this eliminates the suspicion of homosexuality (although hardened homosexuals are never timid before the opposite sex); and the cowboy archly articulates the American male's scarcely expressed conviction that to enjoy easily the company of women—not their favors, please note!—is not quite masculine. This last smacks of Europe, bows, hand kissing and of an overly-glib articulateness.

To those who protest that this picture of the American male reflects the past we can only proffer the still-enormous popularity of the cowboy genre today. This popularity points to the still very vivid male mingling of misogyny with masculinity in our country. Men generally surrender the education of their children to their wives as well as surrendering their bankbooks. They seek liberty even when tied by the marriage knot. They envy the cowboy's heroic, lonely, and very free role even when they feel constrained to sneer at it publicly at cocktail parties. For these reasons the American male invented the cowboy just as Feurbach claimed that man invented God: the one as well as the other reflects what man would be could the heart dictate to reality.

The cowboy is a sturdy holdout in a society dominated by media that scream sexual permissiveness. Whereas Paul Bryan in "Run for Your Life" (with only one year of life to run) can succeed in pleasuring a different woman every Thursday night on television without harming either himself or her, such goings-on are rigidly forbidden Little Joe in Bonanza which immediately follows. The American ambiguity towards the sexual could not be more dramatically expressed and expressed almost simultaneously. But our sexual puritanism is not the dominant note in the cowboy holdover from an older ethic which today is aging but vigorous. What stands out in the cowboy classic is the enormous American respect for law and order. When the French philosopher Etienne Gilson suggested some years ago in an address delivered in the United States that Americans would stop before a red light in the midst of a desert at a crossroads where any driver could see that no other car was in sight for miles on either side, he was accused of making an anti-American statement: he did. The American romance with law and order reaches even to stop lights in deserts. What Gilson did not know, probably, is that most Americans would go through a red light under those circumstances but would do so with the thrilling conviction that they had gotten away with something forbidden.

The cowboy hero always prevents the town mob from lynching the gunslinger. The cowboy always engages in a holding action until the circuit judge arrives on the scene guaranteeing a fair trial. Here the cowboy hero links with the American conviction, expressed earlier, that *any* anti-constitutional or anti-legal act is immoral. You cannot lynch the murderer even though everyone in town saw him shoot down his victim in full daylight. Why do we constantly withdraw recognition and foreign aid from South American dictatorships that have overthrown constitutional regimes? Answer: because everybody in the State Department was raised on cowboy movies where legality equals morality.

The symbol of the cowboy, a national myth, has deep repercussions for the entire classic American approach to the political order. This approach is linked with the American conviction that the political is something essentially written and that what is written not only incarnates but is the Law of the Land. The American approach is anti-Plato although not necessarily anti-Platonic. Plato's *Statesman* insists that government by law is a second-best rule. Personal government by wisdom is superior to rule by law because, although it absorbs the abstract universality of law, wisdom— according to Plato— goes beyond law to reach to the peculiarly existential circumstances surrounding each case. Christianity tried, somewhat uneasily, to absorb the classical tradition of law into the Divine Government of all things by an eternal law: although we must obey the law which is universally valid for everyone, we are judged by its Divine Author who is more intensely personal than we are. Within the American myth as expressed within the figure of the cowboy, the Law takes the place of

Christ. We cannot, therefore, sympathize with regimes abroad that place abstract constitutionality and legality in second or third place. We cannot understand military take-overs in Greece or Argentina because we truly do not and never have experienced any dimension of decency and morality that is trans-legal and trans-constitutional. A military take-over in the United States, a dream often attributed by the Left to the extreme Right, is not only physically impossible in this nation: it is inconceivable and most especially is it inconceivable in the minds of those highly-conservative generals who supposedly would be the instruments of such an action.

Legality will remain a way of life with the United States as long as the nation incarnates its ancient encounter with the Absolute. The American conviction that meaning is never one with personal existence, that subjectivity must gain objectivity, that we become what we are not, that the role bearer is not the role reaches here its apotheosis. We do not trust human wisdom even when we know that it speaks the truth. Were we to do so, we would overthrow the very foundations of American society. If the murderer does not get a fair trial, Justice is violated. This identification of Justice with Legality renders revolutionary action profoundly repugnant to the American spirit. You never, but never, "take the law into your own hands." Ironically enough, the "taking the law into your own hands" is precisely what Plato's philosopher-king, modelled after Socrates, must do. Plato's personalist understanding of politics is deeply offensive to a people that insists that "we are governed by laws and not by men." A moment's reflection reveals the logical fallacy of this statement: all government is by governors, hence by men, who might or might not govern according to laws. The classical tradition divided such men into just governors and unjust, the latter ruling for their own ends and the former for the good of the commonweal. But in all cases government is personal. Nonetheless, logical correctness is quite irrelevant in this context. Even though men do govern according to the Law, the myth insists that it is the Law itself that governs.

The myth imposes itself imperiously upon the cowboy hero who holds up the mob bent on lynching. His very integrity as a man depends on his defending that Law. Even though the Girl who yearns for his hand in marriage be the daughter of the bad Cattle Baron, the Cowboy must sacrifice everything to the meticulous fulfillment of justice executed by a fair trial by jury presided over by a legally-constituted judge. A concession to American optimism is the fact that everything always turns out well in the end, but this last is simply a sop thrown to an audience that demands that everything end well as in the old fairy tales. Nonetheless, the drama must be played out according to the rules of the game. Fair play, invented in England by aristocrats bent upon sports, found a home in America peopled by democrats bent upon justice. Legality has always been a kind of game but only Americans have taken it with such grim seriousness.

It follows, therefore, that the average American has never taken the dictates of philosophy seriously. Our origins in juridical Rome could not be more clearly manifested. Philosophy, understood as a serious meditation on life with consequences commanding the allegiance of the philosopher, has no "rights" that transcend the written law. How could you squeeze philosophy into the Constitution? As long as civilization was understood in terms of Blackstone written down between precious covers that somehow defied the violence of the frontier, the cowboy in myth straddled the street before the saloon—nervous gun at slim hips—the defender of the American way of life. An awkward tall pine stooping tenderly to children—no cowboy ever hit a child; his violence at the service of law; his simplicity mustered within the ranks of the army of civilization; his inarticulateness bowing humbly before the literary; his loneliness put to the ends of domesticity, the cowboy in his freedom—breathing a clean air uncontaminated by an urban world that followed him from New Orleans and St. Louis as he set his eyes and his horse west—guaranteed him the best of all worlds known to the American dream. Chastity without vows, freedom without license, the quixotic defense of the Law even as it sought to suck him into its vortex—the cowboy played out and still plays out his romantic dream in every television room in America before every tired American.

The cowboy—here is his secret—has the best of two very enticing worlds: he lives and enjoys his own while he defends an order of law and civilization which he abandons as he rides out of town, into a distant horizon, forever to the west. He is the eternal image of violence at the service of peace—his sojourn in town ends at High Noon and evil in the dust and the single gun slowly sheathed in holster, sadness on his face before inevitable retribution and death. He is man defending woman. Like all romantic heroes, he demands few favors for the defense. He just shoves on, alone, inarticulate in speech but eloquent in action, always dreaming of a little range, house, wife and children, open to a dream that he must not and truly does not want to make reality. Every American over forty and most over thirty would be the cowboy because these Americans want both of these worlds. Eat your cake and have it too: this is part of the American dream. First Bonanza and Hoss blushing shyly before the latest saloon girl come to town on the stage. Then switch to a Casablanca cocktail lounge with Paul Bryan and a beautiful woman in need who will willingly pay for services rendered. Switch it here and then there—one channel after another. Have it as you will. But that lonely cowboy will ever meander through your heart and he may very well be the best in you.

The cowboy mythology of law and order and of a frontier subdued is not only being challenged by, but is probably succumbing before, a generation that no longer equates morality with legality. The advent of the hippies and of the flower children, the emergence of Black Power, the slogan urging nobody to trust anybody over thirty contrast sharply with the "silent generation" of war veterans who belatedly finished college during the Eisenhower years and who today form the vast majority of those who watch cowboy movies after a hard day at the office. That generation, aging now into its late forties even as it assumes political and economic power throughout the nation, is cut off by a sundering sword not only from the New Left but from millions of young Americans who have already seen everything on television and who have decided to skip childhood, who were nurtured on affluence and who determined early in the game simply to *be* rather than to achieve—why achieve, after all, when achievement itself has become a conundrum? Surrounding themselves with a cluster of shifting symbols often missed even by those older Americans who work hard at trying to keep up with kids for whom work is square, the new generation thus forges a style that is deeply anti-American in the classic sense of the word. An older America, agonizing in streets torn with racial strife and crime, is today dying. The policeman is a hero only to the older generation but he suggests no heroic frontier marshal to the young. In turn, the policeman—as though he were ashamed of his gun-toting role as custodian of the law—attempts to gain acceptance by covering his mission in society with the facade of sociology. He is *really* out there in the inner city to help those kids in the slums and not to keep 'em down.

Our society, while still feeding substantially off the myth of abstract law and order, does not really believe in it in the deeper recesses of its collective psyche. But while not believing in the myth, society does not abandon it because it can find no substitute. Half suspecting that violence of the long hot summers is justified yet all the while hating the anarchy, middle-class America—and that means seventy percent of America—has no new symbol with which to validate its

passionate longing for a return to the days of old when evil was put
down by guns in the service of justice. There is no ambiguity in the respective roles of cowboy and gunslinger but the black and white images blur when a phalanx of cops takes off with billy clubs and tear
gas against looters on Seventh Avenue in Washington. Re-education
in street wars that still lack effective symbolization in the corporate
consciousness has made the nation vaguely suspect that the candidates for high office are playing out roles unfitted for a new era. The
suspicion is justified because that is precisely what the candidates are
doing: playing out roles! They are thoroughly within the old-
fashioned American tradition. McLuhan states that they try too hard
to seem to be what they want to be. In his language national candidates are "high definition" men trying to use the "low definition" medium of television. They cannot pull it off and therefore they do not
ring true to the young. In the vocabulary developed here American
political figures assume a posture: they live up to a standard; they are   *73*
subjects attempting to objectify themselves favorably before the nation; they seek an image. They are admirable instances of the deep rationalist split between symbolized and symbol, meaning and being,
essence and existence. "We try harder" to a younger generation is an
Avis button. Trying harder no longer is the heart of morality. The cool
media, by joining the dichotomy between measurable standards
which are carrots given democratic men who start the race fairly and
squarely with nothing going especially for any one of them, are slowly
forcing Americans under thirty into the acceptance of being, of things
and men as they are, and into the discovery that there is meaning
within being which is utterly other than meaning found in objective
standards which are goals not yet reached.

American youngsters simply cannot work themselves back into
the Calvinist mystique of achievement and into fragmenting their lives
into stages which lead from lesser mechanical achievement to greater.
Moralists often proclaim that this is a breakdown in the ethical fiber
of the nation. No matter how impeccable their legalistic approach to
the good life might be, they have missed the point tragically. Our
youth cannot accept the standards of the past, cannot accept objective
standards at all, because our youth has been bred into a new world
that renders it extremely difficult for the psyche to project itself into
the future and to plan for tomorrow. Projection as a vivid dimension
of human existence depends upon vigorous visual imagination. This
last is the product of a book culture, of a highly-literate civilization.
When we read and when reading is the very substance of our educa-

tion we simultaneously "picture" or visualize content. This structuring of content within the imagination not only permitted previous generations to "see" content that was not depicted visually at all, but it enabled the literate to project backwards and forwards at will. Vicarious experience grew out of man's interior. This was extremely individualistic and it enabled youngsters born and bred in the jungle of automobiles in Detroit to put themselves anywhere in the world and to dream high dreams of splendid adventure. Intensity of imagination went along with, goes along with, high literacy. And in a book-dominated society we distinguish the dull from the bright by their respective capacities to read—hence to visualize—to project. Sharp distinctions between who one is now and what one might become are functions of the visual imagination. The poor boy climbs out of poverty when he can dream a dream, in the words of Martin Luther King. If he cannot read well, he will dream few dreams, and if he reads not at all, he will dream nothing. He will simply be as he is.

74

But the new electronic technology has created a simultaneity of experience that makes the entire globe present to youth. This withers both the capacity and the need to project. Given that this experience is collective, it dulls privacy and the older individualistic dreaming of a future gives way to corporate experience. Our children cannot distinguish between who they are at this moment and what they might be. American youth does not dream dreams. If electronic media have rendered it increasingly difficult to project an imaginative fissure between "being-now" and a future role in life, it follows that these media move towards a destruction of the ethic built around this fissure. This is not a value judgment. Anyone born and bred into the high literary tradition of the West must sense that a new generation weaned on television is impoverished. With such testimony nobody can quarrel because testimony is simply *that* —the statement of a reaction to reality. Nonetheless the new media are forcing the age being born into a return to meaning and intelligibility discovered within being as it is. The young do not want to achieve because they cannot even experience what the desire to achieve might be. The young simply wish to be. The hippie rebellion is little more than an acute manifestation of this insistence. It is utterly irrelevant within the context of this study whether the style of being encountered by the hippies pleases us or not. But it is very important indeed that our society see in this movement the imminent death of an entire way of life built around a high degree of literacy linked with the mechanical rationalism that has dominated the West since the Renaissance. The signs of this are everywhere. Our

children will *not* be objects, will *not* perform for their parents. Therefore their parents perform for them! Their parents become objects,
painfully checking their image every few weeks or months, earnestly
trying to seem to be what youth wishes they were—or what they think
that youth wishes they were. Hence we confront the spate of articles
and books and television programs dedicated to maturity's responsibility to youth. This willingness to desert subjectivity and become an
object for a rebellious generation of children has been rendered possible for the older because they already were accustomed to playing out
the role in every other dimension of life from the political to the economic. But our children cannot do this: they have never been objects
for anyone: they do not know the trick of pulling it off. Punch and
Judy have been reversed. Puppeteers no longer wish to please youngsters but desperately hope that youngsters will accept what they have
to offer. Just as this longing for acceptance in the theater marks the
end of the sawdust trail, failure, so too do the posturings of elders before the young spell out their own defeat before a new world that they
cannot master.

The age of univocal objectification is passing into history. The future may see a tribalistic kind of solipsism in which men simply cradle
themselves in their own subjectivity as they return to a womb-like
state in which all meaning dissolves into a kind of electronic presence.
Or the future may return us to an Age of Analogy. One thing, however, is certain to anyone capable of reading the times: the heyday of
Essence understood as Objective and Univocal Standardization is finished.

This last is perhaps best seen in the political order where it is no
longer an asset for a candidate to have been a poor boy. Anonymous
origins which are subsequently conquered by puritanical striving have
ceased to be a public advantage in the United States. America— excepting, of course, folk America—no longer looks for an Abe Lincoln.
We are losing our distaste for aristocratic and for personalistic origins,
although this is an irony in the light of the "generation gap." The family has emerged again as a factor in national affairs. The Spanish
scholar of Roman law, Alvaro D'Ors, has argued that a republican society can be either democratic or aristocratic. It will be democratic
when familial origins do not count as a factor in the political structuring of the community. It will be aristocratic when men play a significant part in the affairs of state simply because of who they are,
because they have come out of this or that family. The rise of the
Rockefellers and of the Kennedys is profoundly un-American if

75

America be read in egalitarian and therefore univocalist terms. The nation thought it had done with dynasties with the disappearance of the Virginia dynasty early in its history: Andrew Jackson marked the triumph of democratic West over aristocratic South. Yet the age of the dynasts has returned. These men have national meaning because of who they are. Their medium—their families— *is* their message. The glamour as well as the hatred attaching to them is one with their names. They have not striven to overcome obscure origins but have capitalized and even gloried in origins thought to be illustrious by the populace. Their symbolic significance to the country is one with their being who they are. No deeper schism between old and new could be found.

The mantle descends from John to Bobby and now to Ted because of their position in a blood line. Yet this return of the principle of familial legitimacy is a deep affront to an older America. Egalitarianism did all it could to rub out the legal distinctions between legitimate and illegitimate children on the symbolic grounds that children must start their race through life with an equality before the law that was offended by giving any status to familial *being*. This legal fiction has always been contramanded in fact: the toughness and tenderness of blood, of life, are not convinced by the mythology of egalitarian democracy. There is not a father in America who does not want his son to have a head start. Nonetheless, the myth chattered on as though the family did not exist. Legality was ultimate meaning in existence. Impersonal and even anti-familial, legality was our metaphysics. Therefore the rise of the new dynasts has not gone unchallenged. It has found its enemies, gunmen, apostles of the gospel of anonymity of death for the famous and of fame for the obscure. Mass media confers immortality. The mantle descends through blood and murder in symbolic assassinations shaking the nation in its living room as it watched the funeral limousines head for Arlington. Even more: these acts were symbolic murders in the minds of the men who pulled the triggers. Sirhan Sirhan had nothing personal against Robert Kennedy: he had not been oppressed or wronged by him; but the senator bore an illustrious name blending with an America allied with Israel against Sirhan's own Jordan. Pulling the trigger in Los Angeles, he symbolically murdered a nation oppressing his own people. And he shall be followed by more symbolic murderers who clean their weapons nightly, waiting for the moment to gain world fame through ritual killings that will be present mythologically and simultaneously to every corner of the globe through the new technology. All of this bespeaks a profound shudder within an America no longer knowing its own meaning as its keel strains under one of those vast sea changes of the spirit ushering out the old and portending a future still shrouded in darkness over the horizon.

Work and Leisure

What This Means in Education - Politics -
Religion - Privacy - The Kennedys
What the Hippies Are *Really* Saying

THE New Left hates machines. This is as true of Arlo Guthrie of "Alice's Restaurant" fame and of the gentle flower children for whom he sings at the Newport Folk Festival as it is of the hippies who battled the police in Chicago during the Democratic Convention of 1968. It is no surprise that there are traitors to industrialized civilization within the western world. Its history could be written around that very theme. But in the past these enemies of the mechanized have been men "of the right." The surprise of our time has been the swift and unforseen rise of a new leftist international—bearded, defiantly "drop out"—that unites in a loathing that not only sweeps within its hate the physical presence of machines in our midst but that also englobes the standardized and mechanical organization of life patterned after rhythms which were first imposed within the factories of Birmingham and Leeds. An enormous shaggy army marching under placards proclaiming peace and love stirs through the cities of the world. It threatens the established order in Berlin, Berkeley, New York, Mexico City, and Paris. A tool of Communism in the West, the beards and bards threaten Communism in Prague. These latter day Luddites menace a civilization that once found it easy to hang upon a tree the first Captain Ludd who led the machine-breakers in the north of England a hundred-and-fifty years ago. But today the soul of Captain Ludd quickens a movement whose goal is nothing less than the complete destruction of the entire world into which every reader of these pages was born. Captain Ludd was as reactionary as was Cobbett and even more radical. His ghost fills, as would a great wind, a new revolution so radical that it makes conservatives of Communists.

The most distinguished philosopher of the New Left is Herbert Marcuse. Students of the meaning of work and leisure in our transitional age might very well turn to his thought, not only for whatever intrinsic value it possesses, but because it reflects the rage of an entire generation of children who have said "no" to the puritan gospel of work and to the "one-dimensional" society which treats men as though they were machines. Marcuse's *Eros and Civilization* takes its point of departure from the author's conviction that the neo-Freudians have betrayed the original insights of their master. Erich Fromm, Karen Horney, and Harry Stack Sullivan drained Freud's thesis of intrinsic interest by introducing cultural and interpersonal relationships in their evaluation of the human condition. Freud's emphasis upon the erotic, the pleasure principle, was watered down by considerations which are banal when true and baneful when false. Marcuse believes, with Freud, that there truly is a perpetual war between a pleasure principle repressed and disciplined by a production *79* principle. The pot of passion is always boiling over within man and the lid of repression is always keeping it down. Quoting Freud to the effect that civilization is the result of sexual repression by the principle of production, Marcuse parts company with the conviction of the founder of psychoanalysis that civilization is higher in proportion to the lowering of pleasure by the agents of production. Freud was right for the past but not for the future. Marcuse suggests that tomorrow will be built around a civilization grounded in pleasure, a world giving full play to sexuality taken in a "non-genital" but in the broadest and richest sense of the term. Work, hitherto an instrument of repression, will be released through being transfigured by eroticism. Only a new Fall of Man, a second Original Sin, will save us. Work must cease to alienate and objectify. Technological depersonalization—a consequence of the rationalization of work—is bitterly criticized by the aged high priest of the New Youth. He invites mankind to enter, as did Pinocchio, the portals of Pleasure Isle. Marcuse is The Fox leading the wooden boy out of the cozy but corny and confining cottage of the Watch Maker, promising him the world. The World of Clocks and of Duty to Work is abandoned. Descartes is rejected. The Second Adam will be no sorrowful and suffering God who turns wooden men into donkeys because they dare to aspire to flesh and blood. The New Adam will inhabit a Paradise in which men release themselves, in joy and relief, from an immemorial history of technological repression and religious fanaticism as they embrace an Eros stripped of sin and shame. Odysseus is undone. The primordial father is resurrected by the sons who murdered him. Genesis is played backwards. Fair is foul and foul is fair.

Anthony Dolan, folksinger, has called the flower faithful who gather at Newport to hear Phil Ochs and Joan Baez, "sensitive, hurt, rootless, despairing, ignorant—appallingly, vulgarly ignorant." He could not apply these last pejorative adjectives to their intellectual hero, Herbert Marcuse. His erudition liberates him from the charge of ignorance but his own anti-Christian fanaticism constrains him to operate within a frame of reference so narrow that the blinkers limiting his vision look alarmingly like Freudian censors put on by the professor in order that they might repress his own scholarship.

Marcuse locates the perpetual war between pleasure and production in a mythological fall of man which played out its bloody drama in the most remote and dim origins of mankind. We refrain from commenting extensively upon this interesting Freudian hypothesis because it belongs to theology just as much as does the Christian teaching on original sin. Therefore we leave the issue to theologians. Abandoning these speculations on whether our primordial ancestors first fell into Sin or into Work and returning to what is historically verifiable, we insist that the mystique of pure production is by no means as venerable as the intellectuals of the New Left presume it to be. Western mankind bound itself over to the worship of work only a scant four-hundred years ago.

The genealogy of the western mind since the collapse of the Middle Ages can be read in more than one valid way, but there is a lineage leading from John Calvin and the puritan ethic to the ape which is its rejection in Haight-Ashbury. Tawney and Weber and Sombart have traced the history of the work ethic but here we are interested largely in its psychological and philosophical depths. With puritanism the unity of all things within the bosom of a traditional or "tribalized" life was shattered in the West. Power fell to the "godded men" who looked upon politics as an instrument honed to the task of building here below a golden world within which the Lord would dwell in the midst of His saints. The double revelation of God to man, one supernatural and the other natural— insisted upon by the older order—was discarded. Nature no longer had any symbolic value leading man to the Lord. Nature, as a tissue of divine symbols, dissolved. God—"The Totally Other"—retreated to the lonely splendor of his transcendent majesty.

Marcuse insists that the Marian cult was a weapon in the service of both sexual repression and the unrestricted play of the pleasure principle. But the England of Falstaff and of sherry sac, of caroling and of the bawdy Wife of Bath, of Shakespeare's "chopcherry, chopcherry," looks enormously uninhibited in comparison to the America of the grey-flannel suit and the attaché case. The rejection of matter's intrinsic goodness followed upon the suppression of the Marian cult. Mary—principle of darkness and mediation, the perpetual woman— offended the need puritanism felt to approach a masculine God directly. A womanless universe became the raw material of Manchesterianism, good only to be exploited and hammered into use. Thus a terrible psychic tension between a loathing for matter, linked with a feverish engagement with it in the mass production of industrialized goods, filled the puritan world with a restlessness and a burning energy that opened up *new* worlds less grossly connected with this one. The machine, as McLuhan suggests, became the bride of man.

The protest against the dominance of the production principle and of salvation sealed symbolically in work was the baroque, the spirit of the Counter Reformation and of Trent. The baroque was thrown up in protest against the rising commercial spirit of Holland and England. Fused under the force of battle, the baroque was an artificial culture, a theatrical world affirming flamboyantly the glory and grandeur of the things God had made. This was the age of the closed carriage and of the silver cane, of churches as gay as ballrooms. The culture was erotic and it lavished its substance recklessly and spendidly on God and Church. As Dawson has written, the baroque "lives in and for the triumphant moment of creative ecstasy. It will have all or nothing. 'All for love and the world well lost' . . . *'Nada, nada, nada.* " '

The turning back of Alexander Farnese from Paris was one of the crucial moments of history. The defeat of the Castilian infantry marked the victory of the closed over the open spirit, of the cautious over the ecstatic, of the prudent over the lover. Thus the age of great love affairs was brought to an end. The Praise God Barebones won the world and the future was given over to . . . Benjamin Franklin and the idea that a penny saved is a penny earned. The baroque was a culture of the court that enshrined the *Sanctum Sanctissimum* on a high altar as befitted a great king. Stamping forever in its image the Hispanic world and South Central Europe, the baroque burnt itself out politically in a madness that remembers Quixote. The risings in La Vendée in 1793, in Tirol in 1809, and in the Basque provinces in the nineteenth century were the last sparks of the Counter Reformation spirit. American Catholic intellectuals have largely rejected the ba-

81

roque age and this indicates the total triumph of the puritan ethic within the Anglo-American world. But in the very moment when American Catholicism made its peace with Anglo-American puritanism, the latter had its props pulled out from under itself by a massive rejection of all of its values by its own youth.

The revolution of the New Left in our society is not a dropping out of civilization as such. The rebellion is a reaction, not against work necessarily, but against a particular kind of culture dominated by a specific and by no means universal worship of work: work as the peculiarly unique road to virtue and to salvation. That the attack against technologized work and industrialization is mounted today from the Left with a practical efficacy never matched by the traditionalist critique of machine culture is a sign that an old order of things is today passing out of history. Fifteen years ago, even ten years ago, it was impossible to publish an article in the United Stated attacking technology without being branded a romantic conservative. Today a reasoned defense of the technological order is a sign that its author is "square." A new climate has settled into the American soil. The clash of the two technologies within the corporate psyche imperatively demands an ontology of work or a rational penetration of its structure. The repressions that mechanized labor press down upon the brow of mankind cannot be understood unless the very structure of work is illuminated by the inquiring intelligence.

What is work anyhow? Guided by Aristotle, we begin simply by pointing out that work belongs to that kind of activity which the Stagirite called transitive. Today we call it transcending. This activity abandons its origin and settles into an "other" which is acted upon and ennobled. The total perfection of the act of building a chair is the chair built. The entire dynamism of the act of operating surgically upon a sick person is in the patient. Were the perfection of making a chair to remain in the carpenter, the chair would never get built. Were the perfection of the surgeon's skill to remain in him, the patient would die. Work manifests what Yves Simon called a spontaneous and even unwilled generosity. A work act, in the most grumbling and bitter of laborers, succeeds in refashioning the world and it does this altogether in independence of the personal intentions which motivate the worker. Even that classical paradigm of Marxist theory, the proletariat, alienated from the fruit of his own production, works his meaning into the heart of the world which is thereby transformed and elevated. Let us define work as disinterested activity, pure process transcending its own dynamism, service, the generosity of existence.

82

The classical tradition contrasted work with leisure and discovered the heart of leisure in contemplation, the apex of life: immanence. Living acts do not transcend their origins but bend back upon themselves in a circular self-perfection whose dynamism is an intensity rather than an externalization. Loving, knowing, imagining, thinking— these acts perfect the agent acting in an internal exercise of being. If work be process in the service of the world, then contemplative leisure is activity in the service of man. It was upon the basis of these philosophical considerations that the Liberal Arts tradition, subordinated to theology in the Middle Ages because it was nothing more than a scaffolding studied by the very young, emerged in the Renaissance to form a new humanism. This humanism was marked by a contempt for the mechanical arts and for the new mathematical physics. Humanism was a reaction back to the classical and pagan disdain for manual labor. The monastic tradition fused work with contemplation in the service of God, the *Opus Dei*. But the new Humanism created a fissure between the "two cultures"—the technological and the humanistic. The West today is suffering the final consequences of the divorce as scientifically-educated men cannot talk with men formed in the humanities. General De Gaulle's projected dismantling of the famous French *lycée* in the name of technical progress was merely one last instance of the decay of humanism in this winter of a very old world.

83

The opposition between humanistic leisure and pure production was heightened by the rationalization of work at the hands of Cartesian mechanization. Machines would be generous if they could think. They would be archetypes of heroism and would shame the most disinterested of men. A machine with consciousness would discover that it had a heart and would know that its whole life, from inception in the mind of its human maker to final obsolescence and the scrap heap, is consumed in one long repetition of acts of pure generosity. Machines, as we have insisted, are not realities in their own right. They cannot luxuriate in the pure delight of simply being. Structurally processes, all machines exhaust themselves in the work they perform. Comic books that portray mechanical robots as "good guys" and their human masters as "bad guys" reflect our awareness that no mechanism—even if infused with consciousness by some secular miracle— could ever be an egoist. This altruism, built into the very structure of mechanics, simply means that a machine exists in order to wear out, to exhaust itself in doing and in laboring. Machines, pure performance, are archetypes of Work. Work is for what it does. Were work to cease *to do* and pretend *to be,* it would stop being in the very moment that it was.

There may be an incidental aesthetic pleasure in simply looking at a splendid gun collection, for example. This delight is subordinate to our anticipation of how the guns will perform. Work, both human and mechanical, is always linked to anticipation, to a future. It follows that work is the more perfect in proportion to our ability to predict accurately anticipated results. We can do this much better with machines than with men. An office manager can be far more certain of how well his typewriters will work and how long they will last than he can be of his typists. The more rationalized work becomes and the less dependent it is on the human factor the better it can be projected into the future and manipulated. Work truly comes into its plenitude as it is purged of the idiosyncratic and of the purely human. Mechanized labor gives us models of prediction based upon quantitative repeatability. This meat grinder will slice baloney into so many pieces of such and such thickness. Could the world not be treated in this merely quantitative fashion, mechanical industrialization would never have happened. Alice's Wonderland is the only *patria* that could never have fathered a factory.

84

The construction of mechanical models depends upon abstract thinking which purges from the screen of the consciousness every factor irrelevant to the end pursued. Whatever cannot be mechanized or pressed into the service of work to be done is ignored as beside the point if harmless and is suppressed if obstructive to the projected goal. That this particular chair is a fine specimen of craftsmanship is beside the point if I am bent upon using it as firewood. That this gabled mansion is an historic landmark is inconsequential if it gets in the way of urban renewal. In a word: rationalist mechanical thinking aiming at uniform results eliminates the *specific in thought* in favor of the *generic in thought* in order to achieve the *specific in reality*. Attention given the materials at the workman's disposal not aimed at the unique result he intends are not only unimportant: they are positively dangerous to what he sets out to accomplish. Non-rationalized and non-mechanized considerations within pure work are accidental. They must be treated as non-existent. Mechanical thinking involves a dialectic of degradation in which every facet of reality escaping quantification and subsequent predictability is suppressed. The most terse statement of this conviction was Leon Brunschvicg's "whatever cannot be reduced to reason is either non-existent or inconsequential." Mechanical work is the triumph of linear thinking moving through a rational sequence terminating in a predicted result.

Work has always seemed to be more opposed to play than to creative activities which often involve an enormous amount of hard effort, work. Loafing, a terror to Anglo-Americans, is a form of play. When done well, as by Latin Americans and Spaniards, it takes up a good third of the waking day. The Hispanic genius for the Great Three-Toed Sloth threw them out of the race for political power in recent centuries. Play, like loafing, is hostile to the production principle which, according to Freud and Marcuse, is the perpetual enemy of pure pleasure. Play is a poor man's contemplation. The basis of culture, as Huizinga demonstrated, play, is simply activity done because men like to do it. Whereas moments of work enter most playful acts, they are always subordinated to play itself. The child builds a sand castle on the beach because he delights in making it. The heart of play is sheer delight in just being and in being let alone, so that Being Be! Making love is an archetype of play unless debased by a concentration upon specific genital release. Love-play then converts itself into work and the anxiety quotient of American males over forty goes up when the planned results fall off. Impotence is often the consequence of turning what should be good fun into work having foreseeable and predictable results. An element of play can be infused into non-rationalized work. This is typical of good hand craftsmanship annealed in the personality of an artisan expressing himself in what he does. There is a proportion between subjectivity and objectivity, and play and work. The more the subject is involved in what he does as a man of flesh and blood, the more present is the playful. The more personal subjectivity is purged in the service of uniform objectivity the more dominant is pure work. Abstraction is one with work's perfection.

85

Mechanization increasingly divorced work from subjectivity and hence from playful personal fulfillment. Filling civilization from the late seventeenth century until today, mechanization divorced play from work and subjective from objective. Play was restricted to greybeards in the winter of life playing chess in public parks and to children fenced off in public playgrounds administered by recreation programs dominated by the work principle. Sports are instances of play, or they once were. But no longer. Taken up today for reasons of health or social convention or the school's prestige, sports have become instances of rationalized work. Mechanical performance is always measured by univocal standards or models; athletic performance is evaluated by a fiercely competitive spirit reflecting the standardization of machine culture.

Work is among the more ambivalent aspects of human existence.

It can be utterly disinterested and thus represent a high form of heroism as in the case of medical missionaries spending their lives in leprosaria. But work can also depersonalize savagely by repressing free and spontaneous movements within the subjective densities of the person. This is well illustrated by work as time. Work is future-oriented because it projects an end transcending itself. Work, howsoever noble or perverted its goal, must run ahead because process is never being. Any delight taken in the present moment, even that of an archer trying the timber of bow before fitting arrow to wood, is impossible if the entire dynamism of his act is projected forward towards its result. The man lost totally in work exhausts himself. He is well symbolized, as Joseph Pieper states, by a creased brow and a strained and restless mouth. His end is exhaustion because he burns himself out in the service of what he does. It is indeed ironic that this destructive drive to work is a goal for Marxism but a reality for the American corporation which literally breaks the hearts of its best men, tears up their insides with ulcers, corrodes their nerves, and denies them the repose which comes from simply doing nothing at all but doing it with grace and dignity. These men—and they run our nation —are ignorant of joys which are commonplaces for men less ambitious and less conscientious. Paradoxically leisure no longer belongs to the elite, but to the working lower classes. An aristocratic society of the past possessed a sliding scale. Work increased as men descended the social ladder. Today the situation is reversed. No one works harder in our society than the corporation executive, the highly-trained scientist, the manipulator of public opinion, the politician, the man in the upper-income brackets. And no one works harder than does a member of this elite when he is supposedly resting. Everything he does from rising in the morning to collapsing in bed at night is calculated and charted and planned from conference table to cocktail party. His need to get ahead cripples and exhausts. Who can simply remain where he is and *be?* You cannot, insists the saying, stand still without falling behind.

A laicist saint worshipping before the altar of achievement, the American executive forms part of an aristocracy which spreads leisure to the classes below while all the while denying its benefits to itself. The benefits accrued in the nineteenth century of burgeoning industrialism to the repositories of power trickled downwards in a stream which thinned out and disappeared somewhere within the lower-middle classes. We need only read Cobbett's *Rural Rides* to discover how industrialism robbed the new proletariat of whatever leisure it once possessed when it was a free yeomanry. The poor knew no leisure at all, and the Reverend John Malthus wrung his hands in despair over the expanding population growth of the impoverished without realizing that bed is the last pleasure of men denied board. Today the situation is reversed. The captains of the managerial revolution are driven by an inner fear of leisure and by an incapacity to luxuriate in being. A well-disguised and therefore thoroughly effective puritanism haunts our fashionable suburbanites who sense dumbly and obscurely that there is something positively immoral in not working. Leisure as unplanned and spontaneous as Alice's "unbirthday party" demoralizes a class that can be spontaneous only when drugged with drink. The highly mobile and sophisticated classes running America drink altogether too much and drink compulsively. This is a sign of impending exhaustion. Only whisky can slow down a nervous system sawed to the edge of collapse.

Civic activities are marshalled and drummed exactly the way squads execute short-order drill. Even religion is charted with an eye to predictable results. Our elite classes are forced by the imperiousness of a quasi-mystical dictate to plan and organize, to look ahead, to fill time, to project, to put money and land and talent to work. This fear of enjoying life, of *Gemütlichkeit* and of the cheerfulness which comes from accepting the miracle of simply being at all, is well illustrated by the reluctance to retire of men belonging to the upper-middle classes. Retirement means hunting and fishing and hence going to seed, dying before one's time. Now hunting and fishing—the examples are typical and hence symbolic—are *enjoyed* by those who return, summer after summer, to field and stream. But this "enjoyment," genuine play, has been peripheral within forty years of keeping to the grindstone. It has not been central and woven easily and gracefully into the world of work, into what "life is all about." Therefore the American of sixty put out to pasture even with a fat retirement and a bundle of stocks guaranteeing his last twenty years against a great depression that still haunts the attic of his imagination can no longer "enjoy" what he "enjoyed" for years—hunting and fishing. He cannot make central to his being activities which were once peripheral anymore than anybody can make ice cream take the place of steak at dinner. Play which is "ice cream" destroys itself. It is not serious. We will return to the issue.

In any event, the two technologies, for once acting in unison in what will certainly be their last alliance, are forcing leisure on a society bred in the puritan mystique of work. We move today inexorably into a future in which men will work less and less in the older conventional sense of the term. It is by no means impossible that the Anglo-American world will lose its preeminence and that power will pass to Latin civilization because of its innate talent for handling leisure and for weaving it into the very center of the fabric of life. What will happen to Americans who can no longer go-go because they are forced to rest-rest as was the man commanded in Ecclesiastes? *"Et hora furgendi non te trices: praecurre autem in domum tuam, et illic avocare, et illic lude."* When the time comes for going, do not linger; get thee gone speedily to thy home, there to divert thyself and to play (32:15). It is revealing that the author of Ecclesiastes urges play upon us when we are menaced by an impending disaster. What will happen to men who can no longer *do* but who are forced, willy-nilly, *to be even in their very doing?* Some of them would rather die than face being. Ellenberg, in Rollo May's anthology, *Existence,* relates the case of an anxiety-ridden woman who was able to laugh with the joy of living

through only one full day of her life, her last, only after deciding that the day would close that night with her suicide. Existential guilt does not come from a sin which is an act but from the sin which is being. This is the final triumph of process over life. This is why the businessman turned over to hunting and fishing when sixty dies at sixty-one. John Calvin, as he makes his last curtain call on the stage of history, carries his followers with him.

Non-rationalized work depending upon the craftsman's personal skill is superior to mass-produced and anonymous production in that it infuses creative play into the artifact, thus opening the door to delight and free expression. Nonetheless all the advantages of craftsmanship do not lie on the side of subjectivity. Some are objective and look to the product. A hand-made watch or a custom-tailored suit is superior to its machine-produced equivalent. The Chrysler LeBaron "Imperial" is a luxury limousine lovingly brought into being by artists who personally build and install in the automobile a bar, television, telephone, radio, extra foldable rear seats, and similar toys: the LeBaron is a superior product, no doubt about it: five of them were made in 1968 and they cost roughly some $16,000 each. These facts and others like them establish the following set of proportions: although inferior work controlled directly by the skill, in this case inferior, of the artisan is surpassed by rationalized and mechanized mass production, superior artisanry is preferable to the mass-produced product. But there is a reverse side to the same coin: hand craftsmanship, because it defies the laws of quantitative repetition, is limited numerically. The equipment needed to run one first-rate clinic in this nation, if made by craftsmen, would need a small army of intensely-skilled artists engaged over a very long period of time. Rationalized work, handled by machines tended by fewer men, all infinitely less skilled, produces equipment for a hundred clinics.

The problem cannot be resolved within the context of machine culture. Work is either in the service of worker or work. When the former, the quantitative results suffer by being diminished numerically; when the latter, worker sacrifices personality to job. The tragedy is inbuilt and there is no way out. Work, intrinsic generosity, exists for man but it does so, *save* the qualifications outlined, at the expense of the worker. The less personalized delight he takes in what he does the more he serves the community. Pure work within our rationalized civilization makes life comfortable, removes contingency, eliminates pain. The price paid for personalizing the world in this fashion is the depersonalizing of the worker who does it. The problem of leisure within machine culture is acutely moral. Men cannot—must not—express their personalities within the world of work. Leisure in the narrow sense of "free time" and an ever-decreasing work week became an imperative for politicians bent upon procuring a maximum of free expression for human creativity denied expression in factory, shop, and office. The intention of these politicians, some of them statesmen, failed to account for something machine culture might have taught them: no driver shifts gears at very high speed; should he be so wildly foolish, he will face a not-inconsiderable repair bill, and he might even get himself killed. Leisure as "free time" from the slavery of machine tending does not "carry-over" from pure work. Play, in an ever-increasingly affluent blue-collar class, is trivial, non-serious, hence demoralizing. Our elites work even when they play; our working classes—composed largely of minority groups in the north and of depressed pockets of Wasps in the south—play badly in their leisure because their play is divorced from their work. It does not affect their pocketbooks. It diverts from the support of their families. Peripheral, play belongs to week-end sprees. This play is not integrated into the going culture. It is not serious.

For serious play we must go, for instance, to Spain where the *corrida* is invested with the high drama of life, as man faces bull in a duel where defeat is death. Such play, heightened by risks reaching to recesses of a soul encountering bravery and its mystery, tempers and anneals. Serious play is a game. Wrenching men out of daily routine where contingency is conquered and security guaranteed by the monthly premium paid to Prudential, serious play returns to the

89

springs of being. Were men conscious of all this, however, their play would be poisoned at its source. Playing man delights simply in what he does and most especially he delights in risk. This is as true of the splendid baroque cape work of a Manolette as it is of an arm at tiller meeting a head sea blocking safe landfall. What was work for one age becomes play for another. No longer dependent on nature *by nature,* we make ourselves dependent out of a liberty that demands contingency as somehow consubstantial with manhood. Men stalk deer today with bow and arrow because they want to, not because they have to. And whatever meaning there be in the play is discovered in the doing. Symbolized is symbol. Style is being. But play is not restricted to males escaping technological predictability. Are not women bored with men caught too easily? The game element in play keeps it serious and therefore playful.

G.K. Chesterton can help us here: the opposite of serious is not playful: it is non-serious. The opposite to creativity is neither play nor work: it is non-creativity. Men can be creative in both work and play. Non-creativity in work—the *serious* dimension of life within a puritan culture—blocks effective creativity in play. It follows that invading leisure produces lassitude and boredom. Both cause a vacuum opening the door to violence, man's response to meaningless existence. Walk through a Negro neighborhood in Oakland, California. The old wooden Victorian houses stand up stiff and erect, side by side. Hundreds of folk, block after block, sit there on porches, catching the cool of early evening. They say little to one another. They watch and wait. There is silence. There is no peace. But this silence portends and promises violence.

Go into a dance hall in any border state or in the southwest. Listen to the pounding of the juke box on a Saturday night. Note the fake folksiness of boots and cowboy hats on truck drivers who never sat on a horse in their lives. Listen to the band refashion in music an agrarian southern paradise spangled all about with the Stars and Bars and full of an honor unknown but passionately desired. Violence heavy at the dance-hall and the cop at the door. They attest to the loneliness and meaninglessness, to men and women robbed by mechanical work of any capacity to be. On Saturday night there is always a fight or two over . . . nothing at all! Truly, there is nothing else to do. Within— white against white, outside—white against black.

The racial war has tortured roots throughout the nation whose full exploration would demand a library of books. One of its causes is the imperative *to be* forced on men who cannot handle leisure because their work is so deadly boring. The only significant difference left them in a world without other differences is black and white. This is a last vicious vindication of quality's right to exist within an iron world governed by monolithically quantitative laws. It is as though quality said to our world: if you will not recognize that I am because you destroyed older hierarchies built around me, I will return to plague you in the only way I know and I will tell you that black is black and white is white and no amount of levelling will ever wipe out the difference! Your mechanized world is bored about everything else but it is not, nor will it be, bored about *that!* Racism, black and white, is a last stand of personality denied effective symbolization in every other dimension of life. A racial war threatens to ruin our nation but it is possibly the only expression still permitted subjectivity by men who are otherwise objects.

Marcuse grants that industrialized workers *can* find satisfaction in the work they do. He evidences the typist who takes pride in turning in a neat and clean copy. Nonetheless, insists the philosopher of the New Left, this modest pride that men and women find in a job well done is simply a moment within a repressive act forced upon society by the totalitarianism of the production principle. It is candy fed children denied caviar. Marcuse hopes for a future structured anew in an imaginative and revolutionary gesture which will enthrone the pleasure principle and absorb work into *eros.* The march of his argument, rooted in a meditation on the meaning of *eros,* invites us to return, as he did, to the Greek spirit from whence came the very word. *Eros,* in the pristine springtime of classical wonder, was a synonym for the dynamic, *dynámis,* the emerging of *ousia,* of being, from out of the restless imperfection of nature seeking fullness by an internal imperative. *Eros,* hence, is nature's love of itself, a passion to be fulfilled, to be. But *agápe,* within that same tongue bathed in the origins of western thought suggests a personal love in which the lover seeks nothing for himself but rather an affirmation of the beloved. Play, we have insisted, is a delight in activity exercised in and for itself. Play, therefore, bends back upon subjectivity which is thereby affirmed. But were life itself in all its dimensions to be given over to erotic play, it would follow that man would be robbed of something intrinsic to his full dignity, the need to give himself away in the service of others. Here the Old Left can teach something to the New Left. Marx was right in his conviction that labor need not alienate. Alienation is purged from work to the degree to which it is disinterested, to the extent to which the worker loses himself in his work. Any man incapable of forgetting

91

himself before the object perfected by him in work is reminiscent of the solipsistic psychotic whose refusal to weave baskets is a sign of total withdrawal from reality. An absent-mindedness about *eros* is the condition of *agápe*. Orthodox Freudians supposedly are experts about sex: they ought to know that the condition for self-satisfaction in sexuality is the ability to please the partner in love but that this gratification of self depends upon a trick altogether independent of our personal calculations. The trick: you will be fulfilled only if you don't worry about it! Analogously, a world of men who could not lavish their substance upon reality through work, a world so radically subjective that nothing beyond the self could grip curiosity and command interest would produce so many psychological cripples that our hospitals would have to be quadrupled.

Work heals. Even Marcuse knows this because the number of his publications proves that he works very hard indeed and gets plenty of satisfaction out of what he does. Every man who has experienced sorrow and who knows personal tragedy has sensed the curing and hence erotic beneficence of work. But all wise men know that no amount of urging a bereaved or bored person into work "for his own good" will get him moving. Work done for the sake of therapeutic results becomes "busy work" and this often violates the nature of both work and play. Men must be gripped and wrenched outside of themselves by something that they love. Saint Augustine made this kind of love which binds otherwise isolated men into community, the very foundation of all civic life. Objectivity then heals the wounds of subjectivity. In a word: the conversion of the production into the pleasure principle promises very little pleasure indeed.

The hippies demonstrate this. They work too hard at being different. Their revolt against mechanical standardization and the loss of personality within our rationalized and heavily organized society is an over-reaction. It is an ape of the world that is rejected. The occasional conservative crank with a flair for capes and calabash pipes and an inordinate thirst for brandy and soda plus a propensity to toast The King Over The Water as they do at Yale—and this despite the fact that "The King Over The Water" died in 1766—is far more revolutionary in his style of life than the bearded hippie bending over pot. The type is too easily recognized: long hair, loose fitting clothes, beads, even dirt. The concern of public health officials over possible epidemics in hippie neighborhoods points to a savage refusal by hippiedom to equate cleanliness with godliness. The exaltation in not taking a bath is defined in terms of the bath not taken. The hippie is guilty of the

92

over-kill. In saying "no" when society says "yes" and in saying "yes" when society says "no," he dances to a tune he never wrote. Hippie and yippie and flower culture thus produce a uniformity of life style defying its initial intention. When San Francisco hippies put on a mock burial of one of their own in Haight-Ashbury they signalled their own uneasiness with a sub-culture whose mores today, have hardened into rigid conventions. Pot is "in" for the university New Left at this writing whereas LSD is "out." Mores shift rapidly but they are subjected to highly-defined rules. Failure to keep an ear to the ground will leave you "out" just as rapidly as you got "in." Television exposes the monolithic sameness of hippie demonstrations and of the New Left at large in Mexico City, West Berlin, Berkeley, and even in Prague. Everybody looks like everybody else and even those minimal differences discernible between national types in the West are rubbed away in a paradoxical restoration of standardization by the banner carriers of the revolt against it. The New Left's antipathy for machine culture and for the dominance of mechanized work has failed to create an alternative. Despite certain peripheral gestures towards a return to nature, the new Luddites have not gone back to the farm. They have not followed the example of the socialist communities of the last century. In fact, the economic basis upon which the New Left reposes depends upon machine culture itself. One hippie supports six or seven living in one room but he does so by working in the economy he betrays. Most of the flower children as well as the university students in revolt subsist on the charity of bewildered parents. Week-end "dropouts" in the ranks of the rich are by no means exceptions.

This last fact points up the ambiguity felt by the New Left towards electronic media. The young rebels want all the exposure they can get because they gesture to the world their way of life; they expose "police brutality"; they forward Love everywhere and peace in Vietnam. Nonetheless the intellectuals of the revolution express the fear that *all* technology enslaves. The issue was raised specifically by professors associated with the New Left at a national meeting of the American Political Science Association in 1968. Urging a clean sweep of political science curricula around the country because they are cluttered with statistical and behavioristic studies useful only to manipulate men all the more, the spokesmen for the New Left demanded a return to political philosophy. They desperately want to know what happened when Machiavelli released Pandora's Box by freeing the Prince from moral restraints put upon the growth of technical science at the dawn of the Modern Age. Apparently utterly innocent that they were embracing a traditionalist objection to the going order, these intellectuals have themselves failed to elaborate a critique of mechanical culture. They do not know how to penetrate our society philosophically. The hiatus has left the student revolution perilously at sea.

A new wave of conservatism is sweeping the nation today in the name of Law and Order. This reaction, possessing—as it does—power, and glorying in a national electoral victory, might very well swamp a movement that thus far lives only by embracing a pleasure principle subsidized by the Establishment. No "New Order" can come into being without an economic basis. We do not have to look to Adam Smith and John Locke for a justification of this observation: it is as old as Plato and Aristotle. It is not identified with the Puritan mystique of pure work. This failure by the New Left to penetrate critically the mechanical order it loathes is matched by its inability to confront the meaning of work within the economy of human existence. The hippies and the yippies and the drop-outs and the entire generation of dissatisfied youth must resolve this problem soon. Should they fail to do so, their rebellion will dissolve within whatever going Establishment is in power a decade from this writing, an Establishment —we might add—very possibly of angry men forming a latter day Praetorian Guard shoring up order against an internal collapse threatened by another new generation which today, at age fifteen and ten, is as incomprehensible to young men and women of twenty as they are to *their* elders of forty and more.

The hippie is inside the dialectic of modern civilization because of his standardized attack against standardization. The same, however, cannot be said of the entire spectrum of American youth. Adolescents and even post-adolescents are marked everywhere by a refusal to get in the race their fathers won. Young people, for the first time in American history, are not achievement-oriented. The puritan ethic is dead. American youngsters, born and bred in affluence, by and large are not geared to perform according to older conventional standards. Having come into maturity at a time when the cultural lag has been reduced from fifty years to two or three, this generation rebels everywhere against institutional structures whose continuity mocks a life changing so rapidly that the aged are advised to spare themselves the effort to master its vocabulary. Within six months it will be more archaic than Egyptian hieroglyphics. Nothing lasts except affluence. You cannot whip up the psychology of a poor boy on the make in a society in which the poor boy is most decidedly the exception. The middle classes do not understand why their children do not react as

94

they did in their youth. They fret because they have failed to instill stern frontier responsibility into their cubs. But the very success has rendered _their_ work mystique superfluous. Youth will not perform to get what it already has. Affluence is the only continuity in a dissolving world.

This truth is well illustrated in the marked shift away from religious enthusiasm in the United States. Liturgical practices are as "in" and' "out" as are big name bands. High school children comment on the former with the same cynicism that once marked their observations on the latter. Grim and serious nuns, priests, and ministers preaching "relevance" and the last word in "religion" are simply yawned at by kids who already heard it last year when something else was terribly "meaningful." Squabbles about nuns' clothing and hemlines amuse contemptuous youngsters who have already seen every conceivable style on television from Lady Godiva to total coverage in Saudi Arabia. These youngsters cannot understand what all the fuss is about. They are too young to remember the dark ages of pre-Vatican II and therefore Reform and Renewal are terribly old hat. Ironically and unintentionally they link themselves with their elders over forty. Religious enthusiasm has deserted both the very young and the middle-aged. A lassitude is settling over whole segments of the population that are fed up, not only with the older chartering of the spirit in terms of its saleability, but even more with the newer engineering of the soul through liturgical tinkering leaving nothing spontaneous in worship. Lack of spontaneity is a sign of exhaustion. Both point to a drying up of civilization's roots which, when healthy, are bathed in the spirit of play. Romano Guardini insisted that certain liturgical rites long associated with worship could not be explained from any sequential point of view. They have been _there_ in public prayer simply because men liked them! All ritual partakes of the play principle as does every other aspect of civilization from table manners to clothing. When the delight and joy in just being religious is taken away and men are instructed that religion is a process aimed at very discernible results— "Like G.E., Progress Is Our Most Important Product." proclaims a plaque in the Catholic chancery of Baltimore—religion commences to ape machine technology which can never be but only strive.

Religion, by standardizing and streamlining itself to "meet the times," has streamlined all the joy out of it. Old ladies mumbling aloud over rosaries are clucked at by grim modernizers because apparently they like doing what they are doing. This is the only real sin. Play and style have fled along with joy. Whenever meaning is simply one with being as an action, play is present and reality identifies itself with its own style. Meaning progressively divorced from reality becomes abstract and measurable achievement. By an irony hitherto not explored the enemy of the very being of religion today is "meaningfulness." Play is ruined nowadays because it is not serious enough. Religion is being yawned to death because it is *too* serious. Pomp and circumstance, trumpets and scarlet, alternated by moments of silence and liberty, are the only weapons left the churches to capture children under fifteen moralized to early exhaustion by grim elders of twenty-five, in and out of cassock. They are tired of being told that the churches must become "aware" of what is going on: the entire generation of youth is aware of the miseries of the world because they see them nightly on television. They are tired of church services resembling barn dances: they have watched dozens of barn dances on television. The imposition of one medium upon another, folk dancing on liturgy, simply bores the young. Barn dancing is done so much better on television than in Church. It is no wonder youngsters commence to abandon religion as so much nonsense. Conversation heard recently between two bright little monsters: have you been to the Inner City lately? No, but I'm going next week with a Molotov cocktail.

In every dimension of life, therefore, industrialization has forced the play principle and its ally, disinterested leisure and activity, to retreat before the work principle. This reiterates the tragedy of machine technology. The more the worker abnegates himself before the finality of the machine, the more mankind is served and the less served is the worker. Compensated by a progressively higher standard of living, industrialized man cannot handle the new leisure he has won. His culture is centered around the obsolescent principle of rationalist technology and this fails to prepare him for a computerized tomorrow. Education especially reveals this truth. Our entire academy, from grammar to graduate school, prepares young people for a mechanized world which is on its bed of agony. The severely-disciplined structure of the school has been patterned after a civilization built around process, achievement, measurable standards. Children experience no true leisure from the earliest grades through to the highest degrees. Leisure belongs to a world of play which is peripheral and therefore not serious. Leisure bespeaks a holiday on the beach or the summer vacation. The vacation as an institution is only about a hundred years old. It was invented because the enormous pressure exerted upon the human psyche by mechanized culture demanded that the nervous system be rested from time to time. Fenced off in a time sequence conceived as though it were Newtonian space to be "filled," leisure became a badly needed escape from the serious world of pure process, work. True leisure is impossible not only for blue-collar workers but for our burgeoning school population drawn from all classes. Formal education in America relegated "leisure" to "spare-time." In so doing it implies, and sees to it that its victims infer, that leisure—pleasant though it might be—is tangential to the "serious" business of life. *The educational Establishment is utterly innocent of the truth that everything significant in the future is going to be done by men "at leisure."*

Decision-making and creative insight as well as power will be the prerogatives of men who know how to use leisure, not as a "space-time" patterned after an obsolete physics but as a state of being. Belloc hit it when he wrote: "It is in the character of unwisdom to analyse and to proceed upon the results of analysis: in the character of wisdom to integrate the whole point." But we are leaving The Age of Analysis patterned after mechanical models and rigid deduction, all of which involve very hard "work," whether it be the analysis of a literary text into its component parts or of a chemical compound into its elements. We are entering an Age of Synthesis that will demand a maximum of talent concentrated in men who simply occupy positions in society and who will be paid to synthesize, "to integrate the whole point." Donald Cowan, president of the University of Dallas, has coined the phrase a "positional society" to describe the coming age. The roles of the new aristocracy, compacted then with their very being, will consist in unifying and orchestrating into harmony an enormous amount of information fed out of our computerized technology rather than "worked out" in the heads of analytically-trained experts. The Cartesian fragmentation of life, paralleling that of the intellect, required a type of education that slotted people into an atomized world demanding a division of labor which progressively splintered education into a pullulation of specialties. The Age of Analysis did indeed "proceed upon" this character. Academic subjects were blocked out, fenced in, taken apart, and finally put back together again with the precision of mechanics. Even the humanistic tradition which jealously guarded the Liberal Arts finally found it necessary to justify itself in terms of a supposedly ultimate practicality. The liberally-educated man could do the job better than could the specialist! This was a lie. Obviously, he could not! The humanistic tradition, out of a well-merited inferiority complex aped the scientific order, insisted upon the same mathematical rigidity in the affairs of the spirit that operate quite properly within those of machine technology. The humanities

97

were thoroughly technologized. Straight-A students bound themselves
to the procrustean bed of a savage training which involved their getting through such a plethora of written material that a ten-hour day,
often a fifteen, was par for the course. The top graduating senior in
liberal arts from the University of California in 1968 bitterly complained in a national magazine that his four years of college left him
with ashes in his mouth. Not only had he found no time for any significant human relationships, in a word for friendship, but neither did he
have the time needed to meditate carefully and leisurely upon the content of his studies. Forbidden the luxury of simply seizing upon a subject and exploring it from every angle— *at leisure* —bringing to bear
upon it information gathered from possibly a dozen disparate disciplines, he sensed himself to have been on a treadmill which prohibited
his lingering on anything that might otherwise have captured his imagination. Granting that his reaction might be exhaustion for having
worked so hard, it nonetheless points out a profound malaise within
the academy. Youth today does not rebel specifically against parents
as did youth ten years ago and as youth always does. Youth is rather
in plain rebellion against institutional structures which do not seem
"relevant" (youth's word) to the new world in which they will live
their lives. Why spend hours doing homework getting up material that
can be fed back by computers?

98

The dilemma is heightened when we see it in the light of the life
styles created by the two technologies which are at war with one another today. The older still dominating our schools, looks upon education as though it were a highly-structured and well-oiled process
patterned after machine technology and geared to turning out minds
capable of "proceeding upon the results of analysis." Thus a premium
is put on hard work, high grades, and extreme seriousness. But the
newer style of life, englobing and transcending the older within the simultaneity of electronic media, has created an "unofficial education"
which is so encompassing that it is hardly noticed. Chesterton once
wrote that in order to truly *see* black on white you must contrast it
with white on black. This unofficial education is absorbed in leisure
through television-watching, movie-going, and a hundred other instances of the impact of mass media on daily life. A youngster who
has watched some five-thousand hours of television before reaching
the age of five is already highly educated before he enters the first
grade. Now the reaction of the custodians of the educational establishment to all of this is sufficiently well known that we need merely to
cite it: popular education, so runs the complaint, is trivial and banal;

lacking seriousness, it must be positively counteracted within the school. Granting a modicum of truth in the critique, we must insist that children simply are not taking the complaint against seriousness seriously. They live literally wrapped within an electronic world which so encompasses them that it fashions their very style of being.

Orson Welles did Falstaff's famous eulogy of sherry sac on the Dean Martin show. Making up before the cameras, he described the Merry England of long summer nights and gentle fogs, of Prince Harry and of free yeomen and the long bow. Delicately he led an audience of possibly a million people back into the Old England of a Christmas which was Christ's Mass; the ages dissolved by the time he swung on cape and cap and launched into Falstaff's splendidly tipsy defence of wine, fortified and formidable, cup in right hand fondled and adored with due worship. Watching Welles, a master craftsman, tens of thousands of children were *there,* with Falstaff, in a secular miracle that rubbed out five hundred years and rendered time as simultaneous as television's concentration of space within a box the size of a cracker barrel. Now the point to all of this is that the education going on within the viewer's sensibility and intelligence was wrought altogether without *work* in the rationalist and sequential sense demanded were a youngster to get up Falstaff by tracking down the book in a library, plowing through the proper references, and piecing out—with but indifferent success—the meaning of words that were meant to be one with gesture and mime.

99

Watching the war in Vietnam nightly, having the whole world of space and increasingly that of time synthesized within the "idiot box," the generation bred on free education with an absolute minimum of effort, of work in the sense of process, simply bucks against a school system that generally acts as though it were the custodian of wisdom. Now wisdom, good judgment, is dependent on the information it penetrates. More information is gleaned outside of school than inside, and without "the blood, sweat, and tears" mystique of the old order. It follows that young people increasingly come to their own synthesis about life, their understanding of the ultimate questions, by meditating upon a world of facts largely unrelated to those they get down in books at school. Trained for tasks they never will perform according to modes and media which are aging into obsolescence in a world in which everything changes at a dizzying and vertiginous pace, youth has come to form a new class which puts a premium on what it calls "participation."

But official education has conspired against participation due to its fragmented and departmentalized structures. Not only has the establishment frozen the humanistic tradition into a mosaic of books cut away from life, but it is responsible for the sharp break between "higher" and "popular" culture which produced during the fifties an identity crisis within intellectuals who sensed themselves cut off from "the people." Within an organic and integrated society both cultures blend into unity and it is extremely difficult to determine where the supposedly lower gives way to the supposedly higher. Traditional communities do not permit dissection into categories created by the linear and abstract mind. Reality is fused into too tight a unity for rational analysis: like existence itself such civilizations defy conceptualization. The extent of formal education might be trivial and yet the level of culture high: let us but think of the Andalusia of Lorca and the ritual of the ring. The amount of literature published, the number of students enrolled in university and college, might be low and yet the arts might flourish. Only then there will be arts of pulpit and chase, cuisine and dress, custom and manner, arts which do not separate themselves from the community in order to find their proper "essence" within some atomized and abstracted "department," but which fill all society with poetry and soften the tragedy of life: let us think of the Old South and the Cavalier tradition. Finally, even the existence of sharply-defined classes might ironically mask a fundamental egalitarianism, the equality of all men before Death who is made a guest at every table: let us think of Castile and the sombre dignity of the poor of Spain: let us think of the *santos* of the Indians of our Southwest, of the cottonwood Cristus on the way to Golgotha in a cart, surrounded and escorted in dignity by an army of marching skeletons.

The rationalist mentality of the educational establishment, enamored of books and grades and degrees, ignorant of history and innocent of life's mysteries, insists upon judging social reality, wheresoever it might be and in whatever time, in the light of abstractions fabricated in the image of an analytic reason foreign to the rhythm of being and becoming. Official education sends forth into the world idealistic young men and women who join the Peace Corps, the Alliance for Progress, and a dozen other programs that attempt to give an outlet to a generation no longer moved by the puritan mystique to get ahead and make a million. But even these young men and women, America's unofficial ambassadors abroad, enter into their tasks with a solemnity that insists that history and human affection and the poetry of a people are "weaknesses" which must be overcome by the introduction of mechanical technics which will streamline cumbersome and non-rationalized social structures, develop the underdeveloped, and generally help all peoples by seeing to it that they cease to be themselves.

The issue is further complicated in that official education not only fails to induce any significant intuition into relatively non-industrialized cultures, be they western or not; but it also fails to introduce the student into the submarine scholarship needed to penetrate his own industrialized civilization. Marshall McLuhan, in criticizing the Hutchins-Adler romance with the "Great Books," wrote the following charter of independence for the education of the future:

> This book [i.e., *The Mechanical Bride*]   proposes and illustrates some of the uses of this unofficial education on which Dr. Hutchins turns his back in dismay. That unofficial education is a much more subtle affair than the official article as sponsored by Dr. Hutchins. More important still, it reflects the only native and spontaneous culture in our industrial world. And it is through this native culture, or not at all, that we effect contact with past cultures. For the quality of anybody's relations with the minds of the past is exactly and necessarily determined by the quality of his contemporary insights. Thus the failure to come to grips with the particulars of contemporary existence also becomes a failure to converse with the great minds *via* great books. That is why it can be said that the medievalism of Dr. Hutchins and Professor Adler turns out to be no more than a picturesque version of the academicism that flourishes in every collegiate institution. (p. 44)

101

The man who cannot understand, for example, the misnamed Crown of Charlemagne cannot understand the Middle Ages no matter how good his Latin might be. The man who cannot understand why Jackie Kennedy's marriage to Aristotle Onassis chased the war in Vietnam off the front page for a day and diverted the attention of the nation from Misters Nixon, Humphrey, and Wallace has not been educated in his own time within history. Insight into the symbols and myths stirring the soul of the age are glimpses into man's confrontation with the Absolute. When this insight is trained upon the mythology of one's own civilization— and Mrs. John Kennedy was just such a myth—the subsequent revelation is self-revelation. Like confession for the Catholic, it purges self-delusion. Like depth therapy for the neurotic, it educates even while it cleanses. It uncovers the "why" of action and reaction. Like the "Eureka" of Archimedes rushing naked from the baths of Syracuse, it lays open for inspection meaning hitherto bathed in obscurity. A sign of superficiality is to look down from a great height upon the superficial. We can do this only because our feet are planted on the earth way down there below us. Unofficial education has the virtue of humility. It stoops to conquer. But a man annealed in an exclusively literate and official education has been trained, most probably, in a highly-humanistic culture cut away from technological civilization, possibly even in explicit opposition to its values, but fashioned, nonetheless, around its process-oriented and abstracted mentality. Let it be carefully noted that the authors do not suggest that this is bad. But the authors do insist that it is always very bad indeed when men do not know what they are up to.

The literate response to the symbols of popular culture either translates them back into books or ignores them as trivial. The literate approach to the politics of the Middle Ages is to hunt them down in some text, such as—for example—Aquinas' *Commentary on the Politics of Aristotle*. Such a man would never dream of finding those politics symbolized for him in a Reality, a Crown, a Thing. Yet these politics are *there,* concentrated within the Crown of Charlemagne in the Hofburg in Vienna. Anyone who can read that symbolism and synthesize it into imaginative unity has already penetrated the spirit of medieval Christendom and he has done it with materials far more apt than any written text from the same centuries. The history of nineteenth-century Spain could be written from the popular songs Carlist and Liberal sang as they marched off to three bloody civil wars. Vico wrote the history of Sardinia with no more information than the commercial vocabulary then in use on that island.

The literate response to Jackie Kennedy's marriage ran as follows: wars and elections are more serious businesses than marriages: *ergo*, marriages should be relegated in news coverage to their properly modest place; this marriage was not so relegated; *ergo*, news coverage is not serious. Abstractly understood, the reasoning is impeccable, but it is impeccable reasoning divorced from human reality. It suggests the univocal mold into which western thinking has been forced since the advent of machine technology and linear univocal thinking.

This reasoning is heavily moralistic. Machines, thanks to their inbuilt ontological generosity, suggest the same structuring of activity towards externalized goals which accompanies all moralism. Buttressed thus by the puritan mystique, the resultant mentality—a curious grafting of non-doctrinal Christianity to disinterested technics—had to conspire against the play element within culture, thus blocking effective penetration of its cultural importance as being a mirror of man. John Dewey launched progressive education precisely to counteract this excessively abstract de-humanization of our school system. By insisting that education must be "for life," he tried to crack through structures already obsolescent forty years earlier. Dewey failed. It is generally recognized, and this altogether without prejudice to the intentions motivating Dewey, that the "product" was far more pale than the patient it set out to cure. The removal of the school from the classroom to the playground and the subsequent attack upon "classicism," "academicism," "absolutism," etc. was planned according to the linear thinking dominating the older system. The progressivists tried to program spontaneity and to charter education for life. Attempting thus to pump vitality in a box fabricated by themselves, the progressivists brought forth two generations of intellectually-crippled children. Their children—many of the adults of today—were even more harmed in their sensibilities. Ignorant of history, floating on the surface of the present, largely in reaction against an unknown past mythologized as a hideous spook, progressive education lacked the poetic density and the philosophical delicacy needed to surmount rationalist univocity. Book culture is abstract but not banal. Education for life with its planned play and its "learning by doing" was not only banal but ultimately far more abstract than what it set out to supplant. Can anything be more abstract than measuring performance in the school by the square feet and inches making up the school room? This is the absolute triumph of the mechanical mind within the academy. It is Progressivism betrayed with a vengeance.

103

If we would seek an educational theory that did make a breakthrough, modest though it has been, we might turn to Maria Montessori's experiments in the education of children. Begun as early as 1910 and then applied through an international network of schools, Montessori's insistence that environment is not passive but active looks strangely contemporary. Basically it is another way of stating that the medium is the message, that children are educated not by abstracting them from their surroundings but by giving them the tools needed to understand and master them. Montessori began with specially-constructed materials such as blocks and maps and insisted that little ones revere these materials down to the very feel of the wood that went into them. What she said can be said equally of society's need today to come to grips with environment, not only as society fences off its young in schools but as it is covered over by the "skin" of the new electronic technology that educates spontaneously at leisure and without any recourse to the old work mystique. Montessori approached education as an analogical act which is never the same for any two pupils. Whereas progressivism *let* the student take over on his own, Montessori *led* the student to take over on his own, thus insinuating a reverence for the unique character of his personhood as opposed to the rationalist reduction of education to process. Help the student to be by leading him into his *own* being. Forget models and grades.

The stunning success of an approach that gets children to read by age four and to handle the multiplication table by age five has duped observers into measuring Montessori against traditional school methods. Montessori students perform better; therefore, let us adopt the system. But the results of this method, imposing as they are, were merely by-products in the mind of its founder. Maria Montessori refused to evaluate her own approach by these stunningly successful measurements because she rejected measurements of any kind as educational criteria. These last were simply one with making it possible for a child to live within his own world by becoming himself and actualizing whatever possibilities lie latent within his personality.

Thomas Aquinas would have said that Montessori's method achieves high measurable results without even aiming at them because essences, the models upon which all standards are set, are simply univocal moments within existence which is absolutely different in all men and things although analogically alike. The relation will be evident to students of philosophy but it is historically verifiable as well because Maria Montessori studied Thomistic thought in Rome. Be that as it may, the univocal has dominated the western mind and sensibility since the heavy rationalization of work brought about by Cartesianism. Univocity at work in technics is simply a synonym for mass production in which each instance of a Ford is judged in terms of the master model governing the construction of the individual automobile. Univocity at work in popular culture is the Miss Universe contest in Long Beach which quite properly repudiated the Miss America affair because the latter, under the influence of a delayed puritan reaction, introduced talent and poise and other irrelevant considerations in its choice of feminine pulchritude. Miss Universe makes it to the crown because she measures up to a preestablished standard of beauty—so many pounds and such and such measurements and no nonsense about talent. The Fords are checked out, one by one, for brake, clutch, and steering specifications. The beauty candidates are checked out for bosom, waist, and thighs. Univocity in education gives us a 4.00 grade average, Straight A, the Dean's List, academia's

104

equivalent of a 36-26-36. The Fords, the Queens, and the Scholars have one thing in common: their excellence, in each case, is determined by a preestablished standard projected into the future even before they appear on the stage of present time where they measure or fail to measure up. Measuring up, in turn, involves visualization of the capacity to predict excellence or failure before any situation occurs and then to place it into its appropriate slot in the file cabinet of the mind and sensibility when it finally makes its appearance.

The going system of education in the United States and in the industrialized world at large, built around a univocal and analytical approach to the real, has faithfully reflected a society structured by machine technology. Education in the future will seek a new principle of unity because the old has decayed. This principle cannot be univocal. Univocal unity depends upon linear thinking in which individual instances are marshalled under some common type permitting us to visualize measurable results *before they occur.* Analogical unity, however, is marked by its lack of any predictable *vision.*

105

Let the reader make the following experiment. Imagine himself to be sitting in an outdoor restaurant in the Mexican village of Ajiihic fronting on Lake Chapala. Let there be two couples in the restaurant, forming one unified situation. Outside there is another unified situation. Three boys are splashing about in the reeds and riding out the last ripples of the tide on old tires; another lad passes by on a pony; a tourist woman perches on a rock painting the lake; a peasant puts down a bundle of firewood to light a cigarette. The unity found *within* the restaurant is univocal and easily grasped: couples seated at tables form a type which can be visualized and repeated indefinitely in a linear series of similar recurrences in which couples seat themselves in an indefinite number of restaurants circling the globe. It takes little talent to grasp *that* situation. The unity outside the restaurant, precariously formed by a momentary concatenation of circumstances never to be repeated cannot be pressed into any common type. A drama in being, the unity is grasped once and forever or it is not grasped at all. The things and actions related are *one* because they are absolutely *different* from each other. By a paradox, things and actions are thus united in their existential differences, as Gerald Phelan expressed it. Broadening the analogy we note that relatively non-rationalistic societies possess an analogical unity seized intuitively only by the poetic or philosophical imagination. Beer steins, the Passion Play at Oberammergau, a mad king hiding in a mountain castle at Nymphenberg, the *Hofbrauhaus* — all together spell out Bavaria, schmaltzy to some and adorable to others. Each single element detailed can be captured in visual imagery but the unity making them *to be* Bavaria escapes a single model or type. The unity is grasped by intelligence and sensibility in a judgment in depth, an "inscape," an analogy, a fusion of a host of media at work which are not superimposed but which mirror one another simply by being intensely themselves. The layout of the front page of a daily newspaper requires this knack of seizing upon an analogical unity in the news that cannot be reduced to any overarching *type* of news. Pope Paul VI lands in Bogota and on the same day Russian tanks roll into Prague. The stories are juxtaposed along with photographs in every principal paper in the nation. The unity, in this case one annealed in irony, is immediately caught by the reader who cannot, however, conceptualize what he intuits. Were he to do so, meaning would dissolve in nonsense. Millions today live an analogical life without knowing it. They exercise and play out in their being a principle they could never express in theory. Mass media have forced them out of univocity even though these same millions make their living within a machine structure that abhors analogy as does nature a vacuum. The Lawrence Welk show is to a Goldwater Republican what Peyton Place is to a liberal Democrat: it would take a paragraph to demonstrate the equation but everybody tuned in to media seizes the analogy immediately. Any demonstration, proceeding upon linear and therefore univocal lines, would be an anticlimax. It would be arguing the obvious as did Descartes and Company when they proved the existence of the world.

Analogy implies hierarchy and order. A shoe and a dinner are diverse objects having absolutely nothing in common. But a *good* shoe is what a shoe ought to be and a *good* dinner is what a dinner should be. A youngster switching his television set from one station to another carries out an internal argument with himself about hierarchy. When he finally settles upon this or the other program he establishes some hierarchy even though it is governed by no predictable or projectable model. An older hagiography insisted that sanctity in the fierce and intractable Athanasius—himself *contra mundum*—was not sanctity in the modest and sentimental Teresa of the Little Flower abnegating herself in the world. This hagiography was governed by the principle of analogy which tried, usually with indifferent success, to prevent Christians aspiring to holiness to model themselves after canonized men and women who made it by being themselves rather than carbon copies of anybody else. A very good way to understand the electronic revolution in media would be to read Butler's *Lives of the Saints.* The hierarchy forced upon us by the communications revolution which abolishes space and time is analogical. But as yet only the kids seem to be able to pull it off.

Marshall McLuhan insists that the world is being tribalized once again. He was attacked in the *New York Times* for smuggling the twelfth century back into the twentieth—a grave sin indeed! —and he was damned in *National Review* by Frank Meyer for being the apostle of a new barbarism. Meyer could be right but he could also be very wrong, which indicates that he has got his genera on sideways, itself a blunder for anyone in love with the past four centuries. Barbarism is a form of traditionalism but not every traditionalism is barbaric. But what all traditionalisms have in common is a refusal to conform to the mechanical and atomized world of the past four centuries which Meyer equates with civilization. Any duel about the electronic future is going to be fought because somebody throws down the gauntlet in defense, or in contempt of, the world born with Good Queen Bess in England and with Descartes in France.

107

Possibly there is a confusion underneath this "conservative" reaction against electronic technology as exemplified by Meyer. This is an underground cable linking New Left with Old Right. Viewing electronics merely as an extension of mechanical civilization, both predict an even greater depersonalization in the future. Both miss—in our judgment—the real moral problem: the danger of an *excessive* personalization. Depersonalization was the consequence of abstract atomization and the division of labor, both effects of the increasing perfection of machine technology in the nineteenth century. Personal vision had to be narrowed at that time. The world was heavily specialized. Blinkers were the price paid for progress. This reduction of vision limited personal activity to the immediate family and very close "personal" friends. Charity began at home. Often it stayed there. Life outside was so anonymous that only silly sentimentalism justified sallies into it. It was better for charity to stay home. By a paradox hidden in the mists of tomorrow, individual personality is going to be found, not within the spatial confines of one's den or backyard or neighborhood, but in a public order utterly foreign to the division of life into public and private into which we were born. In this sign of tomorrow we can only be comforted, even while we shed cherished values, by returning to a very remote past. Kings were once public sacrifices to the people. A hooded executioner beheaded Charles I and his blood covered all England in reproach. The nation went up to the block with the Royal Stuart and went down, headless, to the depths of its desecrated conscience. Mass guilt made the Restoration possible.

The shudder that went through England was an act of pain. Nothing is more personal than pain. Each man's pain is his own and his suffering can only be shared analogically by somebody who has also known a valley of tears. This sharing hitherto has been the closeted privilege of husbands and wives, parents and children, who sensed— often dumbly— that it would be indecent to reveal this last secret of the heart to a cold and unfeeling world. We hid our pain with good reason. Nobody "outside" cared enough because nobody outside could experience directly what was going on "inside" hospital walls and bedrooms where men lay dying. We kept silent. Jewels are not cast before swine and pain shared is the ultimate jewel. Anybody outside the secret is a swine! Death and irretrievable loss leave survivors alone and isolated. They cannot communicate with a world that does not understand because it cannot. But it takes little imagination to see that today we are entering an order of things in which every man, tuned in to every other man, will share the pain of the whole world. Christians might find here an analogy to the Pauline teaching about the Cross. But the man of the future, be he a Christian or not, will never again be able to "shut out" the world nor will he be indifferent to its sufferings.

If civilization is equated with the private world of recent centuries, then Frank Meyer and those who think as he does are right in protesting against the new age. If we buy the package, however, then we must kiss goodby to civilization. Privacy and all those properly loved values it nurtured will soon belong to the past. Privacy involves a certain absolutizing of space. Privacy was invented by Isaac Newton. The private man—Mr. Aristotle Onassis on his floating palatial yacht—surveys a spatial world that offends and then he steams away into another, more distant space. Onassis escapes, with Mrs. Onassis, a space filled with prying eyes. Mechanical technology made this kind of escape possible, at least for millionaires. Yachts will take you anywhere. "Nature" is the presupposition of privacy. You can only "beat it" or "run away" within space. Paul Bryan in "Run For Your Life" escapes his fatal illness by plunging into the wilds of Africa and the delights of the Riviera. Always on the go, he is nonetheless watched by us wherever we are. Privacy remained an absolute just as long as nature was absolute. But nature is no longer what it was.

Nature—"Mother Nature" until after World War I—has had an exceedingly bad time of it since the Industrial Revolution. Phased out by bulldozing, bombing, asphalt civilization, the mechanization of sex, the obsolescence of craftsmanship, the removal of materials from artisans, and the tree that just won't grow in Brooklyn, Nature, were she still Earth Mother, would be a lady in dire need of rescue. The melodrama could not be more obvious: New Leftists and New Romanticists banging away at tinny pianos in the pit while Prudence Penny is strapped to the tracks, the cruel mechanical beast hurtling down the rails bent on her destruction. Melodrama, of course, is always bad art because it is such good life, just as corny as the real thing. But Mother Nature on the rails—the heroine in the flickers—is never killed. Always threatened by machine technology, nature is abused and diminished and even perfected but never totally absorbed. Otherwise, machine technology would grind to a halt. It would have no fuel to feed its greed, a greed—let us never forget—in the service of a total generosity. Presumably, the train roaring down the tracks toward the Lady in Distress was set in motion by Man.

But electronic technology neither eats up nature nor pours itself out in the generosity of process. Like Popeye electronic technology simply is. It does not destroy nature but covers it over with a new skin—a "noosphere" as Teilhard de Chardin put it. Chardin, however, implied that the evolutionary process would destroy everything that preceded it. Nature, in his view, would simply cease to exist. This thinking, motivated as it was by the dialectics of Hegel, insisted that nature would be so obliterated by its progressive identification with consciousness that nobody in the Omega Point could go hunting and fishing. Chardin would have been right, of course, if mechanical technology could convert itself into independent existence. We have argued that it cannot. Nature—once Mother Bosom for romantics and Raw Material for utilitarians—has been deflated and therefore saved. Nature can breathe again and relax. One million Americans went skiing this past year and more than a half million own sail boats. Erich Fromm writes stern moralistic books out of a puritanical Freudianism about the courage needed to face an industrialized world without nature and then uses his royalties to buy a house on the most primitive coast in Mexico. It is all very comforting. Nature is here to stay. Our future return to nature has been dictated by our having transcended nature. The violations of four-hundred years are coming to a close but now nature must pay for its newfound repose by ceasing to be Mother. An honored sister, nature is just about where it was in the twelfth century. Humbled, but still reasonably intact, nature has been given a new charter to exist by electronic technology. We are going to be able to look at nature again as transistors and computers get

109

smaller and smaller. This could even prove embarrassing. How about the bugging device that does duty for the olive in the martini?

Privacy is dependent upon old-fashioned nature. Nature's wide open spaces remain a possibility for, let us say, Mr. and Mrs. Onassis. People can still "get away from it all" within natural space. They can run off to Acapulco, but—the issue shall command our future attention—they escape *inside* a nature which is itself *inside* electronic media. Who can doubt this when astronauts photograph the earth: what is inside of what? The shots taken of a round world are *inside* nature in the sense that they are made out of elements drawn from nature. But they are also *outside* nature in the sense that the earth itself is *inside* them. Even the telephone, closer to us in its omnipresence, upsets the best-laid plans of mice and millionaires. For this reason we have the unlisted number, the barrier of secretaries, and—finally—the phone off the hook. This last points up the futility of spatial distance before electronic simultaneity. Anybody with the phone off the hook feels guilty.

The ideal of privacy has had a relatively short life span in history. The Greeks and Romans lived publicly. The forum has been perpetuated in Latin culture in the plaza. Life centers around the *paseo* or afternoon and evening stroll. The town turns out to take the *aperativo* with due solemnity mingled with gaiety. Friends meet in restaurants and cafes. Children pirouette around fountains and statues. The young flirt with elaborate discretion as boys walk one way around the *plaza* and meet the girls coming in the opposite direction. What happens when they meet can be electric. Sweethearts are seen side by side during these sacred hours and they mingle freely with their elders. Their courtship is literally conducted in the open and in no other way. Often it is prolonged for years. The lack of privacy would be intolerable to Americans of any age. Friends are rarely received at home because they are seen in public every day of the year. Moving farther back into time, the residual privacy found in the Latin world dissolves totally. Entire families slept in the same bed in England and in Scandinavia during the thirteenth century. Moving again forward in time, a man's privacy disappeared as he mounted the social ladder. Life was corporate and at the top it was totally public. The French monarchy illustrated this identity of a Public Thing with personal life through its incarnation in a dynasty, a family. King and Queen ate in public, were undressed in public, prayed in public, died in public. They were, in Belloc's words, "sacrificed almost as in a public sacrifice— condemned to the perpetual burden of being mixed into this Idea and of supporting the burden of its intensity and power."

Traditional societies live in the open. The triumph of machine technology with its concommitant rationalization of work produced, along with the division of labor, an ever-increasing divorce of the private from the public. Meaning ceased to be incarnated in families and dynasties. The role, as indicated earlier, was distinguished sharply from its bearer. We need only think of the executive in big industry: he gains his position, a job; it does not repose upon him as a person. The internal logic of rationalism, progressively abstract and impersonal, severed its ties with flesh and blood. The decline of the family in public affairs led to a natural reaction. Cherished all the more privately, the family abandoned its old public role. Privatization was not only forced on personality by the demands of the mechanical order but was dictated by personality. Incapable of adjusting the cloth of humanity to the machine, men found a corner for their peace, their idiosyncracies, and the warmth denied entrance to the world of work. When work cut its older ties with play and leisure, publicity cut its with privacy. Nothing was more human than a Dickensian home in Victorian England. Nothing was more inhuman than Victorian England. The war between familial privacy and public demands was illustrated best by the disappearance of the family from participation in the business of politics in all the West. Where today does the family have the franchise?

The privacy sought and found imperfectly by the lower classes was gained effectively, if precariously, by the rich. The older aristocracy in England lived publicly until the nineteenth century. This was part of its duty to the community it led. But the new industrialism forced the upper classes into a retreat into privacy that became a rout with the advent of photography. Nonetheless, the English nobility, due to lingering links with a landed gentry itself in living contact with a yeomanry; prompted by its public role as custodians of the rights and perquisites of the House of Lords; and, finally, moved by its ancestral sense of duty to the Crown, never totally hid itself from the prying eyes of the nation. But American High Society, vastly richer, possessed of the means needed to achieve utter privacy, without public responsibility, has—until recently—lived the most secretive of lives. High Society in democratic America is the most closed society in the world.

Privatization of life was a natural reaction against standardized industrialization. The anonymity of government in the autumn of the old order following World War I nurtured into bloom the Fascist and

111

Nazi reactions in Italy and Germany. It was a cold world crushed by depression. Tens of millions saw their savings dissolved, homes and farms foreclosed, jobs liquidated, lives disintegrated into ashes. They looked for a scapegoat to blame and there stared back a world without a face. There were no more kings and emperors to take the responsibility. Responsibility had ceased to be personal. Aristocratic families, more powerful than ever, had retreated behind a screen of privacy. Politicians came and went. Their names were forgotten as soon as they were memorized by an embittered world. Fascism proceeded upon the assumption that the world was awry because it was governed by secret groups of anonymous capitalists and occult forces, Jews, conspiring in hidden and dark places as they manipulated industry and finance in such fashion that their power crushed down upon the brows of small and independent men catapulted backwards into the whorling vortex of proletarianism.

112   The secret of Hitler's success was lack of information. Anonymity from above tended to reduce the middle classes to anonymity below. Those men caught between the Marxist and Capitalist nutcracker sought to save themselves by fixing their rage upon men whom they had simply *never seen* and who, in some cases, did not even exist. Symbols did duty for Public Figures and thus drew to themselves the rage and frustration of millions who only wanted to be left alone in domestic tranquility. Upon these fears Hitler and Mussolini rode into power. The Hero returned to the Forum. He was swept to glory by the same lower middle classes whose small shops and farms and businesses were threatened by a faceless world fashioned by mass industrialization.

Had television been popular forty years ago Hitler never would have become chancellor of Germany. The perfection of mass electronic media makes the Fascist protest against atomization and egalitarianism progressively irrelevant. Media dissolve the bases upon which private power operating behind the screen of anonymity is built. The conspiracy theory of history, the heart of Fascism, will disappear. Very soon it will be impossible to discover any conspiracies because there won't be any! Even vote-stealing, that venerable American tradition, is today in for very rough sledding as media cover every corner of the nation and let us all in on the secret. A new light, glaring and unpleasant, dissolves privacy. Howard Hughes can hide but he needs a billion dollars to do so. The old distinction between who a man is in private life and what he does or stands for publicly, between his being as a person and his institutional role, between existence and essence, being and meaning, is fading out of history. Life style and achievement are the same for a generation bred on television. From Miami Jackie Gleason introduces Richard Nixon in Madison Square Garden and away we go! Action is acting and acting is being. It is no wonder that this generation cannot get excited about work. It only wants *to be* because it already *is* enveloped within a whole world that it is a part of.

Its hero is John F. Kennedy.

By older American political criteria The Thousand Days was a failure. The Kennedy Administration did not push its legislative program through Congress. The Bay of Pigs and the Berlin Wall were phantoms haunting the New Frontier. But the Kennedy era had a life style of such brilliance and elegance that it captured the imagination and emotions of the world. The Kennedys were a public people and this return of the family to the forum marked the coming of age of a new politics dominated by mass media. John Kennedy was the first American politician who understood television. The famous initial debate with Nixon sealed the presidency for the relatively unknown senator from Massachusetts. Trained news commentators evaluated Nixon's performance on that fatal night to have been slightly superior to Kennedy's on debating points. But debating points were swamped by the medium. Nixon, haggard with illness and jowly, dressed in a light suit that faded into the background under the glare of the kleig lights, was no match for the carefully made-up and dark-suited young Lochinvar whose message to the nation was himself.

113

Kennedy narrowed almost to the vanishing point the distance between meaning and being. He did nothing for the Catholic Church and was adored by the Catholics of the world for being one. He failed to tear down The Wall of Shame and he won the hearts of every Berliner when he declared before them, *"Ich bin ein Berliner."* He permitted the gallant little band of Cuban exiles to bleed to death on the beaches at the Bahia de Cochinas and their survivors worshipped him in Miami when he publicly embraced their banner. The most intensely pragmatic of men, he captured the allegiance of America's intellectuals and artists. He won Latin America by launching the Alliance for Progress which did very little to alleviate the troubles of that continent but which electrified its millions of impoverished simply because it was. And finally he passed into legend as a martyr not because he willed to be one but because he was one. In John Kennedy meaning became again being, a manner of existence, a style of life. The Thousand Days closed the last Four Hundred Years.

We have argued that public meaning in the past was a measurable commodity evaluated severely in terms of results. Tomorrow both will blend into a public life style. A new rhetoric will soon flower in a society for which acting will be consubstantial with being. The price we must pay is the loss of our privacy. By an irony which is a tragedy, the universal opprobrium heaped on Jackie Kennedy for refusing to live out her years a public sacrifice to the myth fashioned by mass media was well summed up by one news commentator who said that everything could be forgiven her save her passion for privacy. When asked why she married Aristotle Onassis, she placed privacy high on her list of priorities. A billionaire who rose to riches from the dockyards of the Mediterranean, Onassis secured *his* privacy by buying an island which is literally a private principality. One newsman bitterly complained that Onassis had his island rubbed off the map so that survivors from shipwrecks would not disturb him. Be that as it may, he surrounded himself with a private army and he lived on a private yacht. Marshalling his vast fortune and pressing it into a war against a world which has been "turned on" by the glare of electronic technology, Onassis fought a rearguard action against the future. But even his fabled wealth will not save him from the fate suffered by all men today who are *public* figures. In seeking her own privacy, Mrs. Onassis has robbed her second husband of his. That yacht's future meanderings will be charted in every newspaper in the West and it will dock, no matter how remote the landfall, in the living room of every family in the industrialized world. What was it Belloc called the French monarchy of the Bourbons? A Public Sacrifice. What price will we pay for an intensely-personalized future? Our privacy. And it is gravely to be doubted whether most of us living in this transitional age are up to it. But candidates for the presidency of the United States are: Humphrey and Nixon were accompanied everywhere by their families in the campaign of 1968; what the American people buy today when they elect a president is a package deal called the family. Even as the family languishes in crisis at the very fag-end of industrialized civilization, it emerges once again within the new age created by electronics. But not only have families become public once more. Prelates have as well. Cardinal Cushing's defense of Mrs. Kennedy's marriage to Mr. Onassis produced such a storm of protest that the aged gentleman announced his retirement two years before schedule. Privacy tomorrow will be the exclusive prerogative of failure. Power will belong to the Public Man. He will have no skeletons in his closet because the closet will be lighted up along with the rest of the house.

114

Publicity reveals the whole man. Total publicity reveals him totally to a new sensibility or sense ratio itself opened totally by electronic media. Marshall McLuhan's celebrated distinction between hot
and cool media looks to the way in which the senses are engaged in
their participation in environments. Hot media narrow sensorial activity to one or two senses. It follows that personality is proportionately
narrowed. Cool media, on the contrary, both decentralize and integrate sensation analogous to the way existence paradoxically inte-
grates and decentralizes. A man is more than the sum of his parts. If I
grasp him as a whole by engaging the whole of my sensorial equip-
ment in my encounter with him, then I synthesize in the psychological
what is already synthesized in the real order. This is why telephone
conversations are less satisfactory than person to person meetings but
more satisfactory than an exchange of letters. Television, still in its in-
fancy, synthesizes better than any other electronic medium because it
involves the whole personality, especially in intercommunication television where we can talk back. It does this in a cool or relaxed fashion. Hot media, such as radio, seize upon one analytically-fragmented
sense and thus mirror the general fragmentation produced by rationalist culture. Radio "personalities," such as the Reverend Charles
Coughlin in the thirties, are often flops in the flesh—or in the tube.
Hot media centralize in opposition to the decentralization intrinsic to
cool media. The "hot" educate by narrowing as does the eye squinting
down a microscope or along the barrel of a rifle. They demand an intensity of attention and discipline that keeps men keyed up, tense,
nervous. The decentralization of the nervous system, the concommitant to the coming decentralization of life in society at large, was first
emphasized in Maria Montessori's approach to the education of the
child. Aiming at a balanced sense ratio, her method prevents any sen-
sory imbalance caused by pushing one sense at the expense of the others. The ideal, of course, would be to give eyes to the blind and the
sensorial delicacy of the blind to those with sight.

In the rationalist culture we are abandoning today, men were educated sensorially in a most narrow fashion, hence the over-extension
of the eye as an instrument in the service of abstraction and predictability. Rationalist education necessarily places a premium on the ability to generalize and a minimum on the ability to integrate
imaginatively. Imaginative integration of the real is opposed to the
analytic isolation of one sense from another. Synthesis in sensation
leads to synthesis in intelligence whereas analytic sensorial fragmentation goes along with analytic fragmentation within the mind. We need

only contrast, for instance, the skipper of a windjammer with the captain of a motor ship. Both must be able to size up a potentially dangerous nature threatening a watery burial, but the responsibilities of running a motor vessel are child's play in comparison to those of handling a big sailing ship. For this reason the German government insists that merchant marine officer cadets train on sailing ships even though they will never see one again in their lives after receiving their credentials. The Chileans and now the Colombians do the same: therefore the former have the finest navy in South America. The wisdom of these nations is a bow towards the educative function of synthesis. The judgment the sailing master makes is a synthetic marvel fused within a sensibility penetrated by reason: he plays it by ear. It would take thousands of pages to explicate analytically and in literate sequential fashion just why he does what he does in a single voyage. Similarly it takes vastly more time to analyze a boxer's skill in the ring than it does for him to exercise it. The synthetical act in which reality is stiffened into unity finds an analogue in the simultaneity of space and time sprung into being within electronic media. The new technology mimics its human master. Both mimic existence.

Analytic dissection of any synthetic unity proceeds through a linear space taking time, plenty of time: a sportswriter needs more time to dissect into its components the third inning of the World Series of 1968 than it took the Tigers to pile up their impressive cluster of runs. Sensorial integration shading into imaginative and poetic synthesis, penetrated by the unifying activity of the mind, makes up the heart of popular culture everywhere.

A similar situation today is the reaction by the custodians of higher culture to the Beatles. The snobs among the music critics simply reject them as being outrageous and barbaric. Reverse snobs write learnedly of diatonic cross-harmonies and the like. But the Beatles happily announce in their official biography that they are a group of ignorant English moppets having a whale of a good time and getting paid handsomely for it. The analytic mind is constrained by its very earnestness to seek deep and dark meanings in the Beatles and in Twiggy. The questions these butterfly catchers ask popular culture and its heroes are posed in sequential and rational terms. Framed analytically, they aim at penetrating the organically synthetic. It follows that the literate mind often falls into romanticizing the real world by discovering intuitions which are thought to be quasi-mystical simply because they cannot be penetrated by analysis. Whatever is foreign is always mysterious. This works both ways. If the Beatles seem darkly mysterious to the music critics, then the music critics must seem the same to the Beatles. The point is grasped when we see that the analytic and the synthetic minds, as well as the sensibilities in which they are bathed, are radically opposed to one another. So too are the civilizations built around each of them.

Given that the age being born out of the catalyst of history will be synthetic, for reasons argued throughout the pages of this book, it follows that education will be synthetic. What is demanded of us if we are to redeem the future is a higher culture and a higher intellectuality which *analogically* mirrors, and is mirrored by, the spontaneous and popular integration hammered into existence under the pressure of demands which are one with the human thing on the streets and highways of the world. The very concept of "higher" and "lower" culture will dissolve. In an analogical world things and men are simply different. Whatever unity they achieve has nothing to do with their abnegating themselves before somebody else's model of what is culture, high or low. Equal, unequal; high, low—these are Byzantine relics inherited from the past. To the mind and sensibility annealed in analogy every man is both equal and unequal to every other man in a bewildering and indefinite number of ways. Questions about equality have only a peripheral interest to anybody who has transcended machine technology where Fords are equal to other Fords. Whether the older and rapidly dying America of *The Federalist Papers* was dedicated to equality or inequality will be a picturesque and lively problem only to antiquarians. The answer, of course, is utterly beside the point to the world today which could not care less. One is reminded of how many angels can dance on the head of a pin.

The technocrats, archetypes of mechanical efficiency, will not run the show tomorrow because the best technocrats are going to be little boxes called computers which get smaller and less obtrusive every year. Technocracy has been whipped by having succeeded so brilliantly. Depersonalized functions no longer exercised by persons take care of depersonalization by eliminating it. The analytic mind—at its best coldly brilliant, at its worst, fanatical—reiterates a violated nature and a shattered humanity. If analysis forms the moment of pure work within the spirit, then synthesis is its moment of reflection in which man looks back on work done, on information marshalled before the bar of mankind presided over by a judge at leisure.

Called upon to synthesize a world, soon a cosmos, eventually every cosmos onto the last and most distant star, man at leisure will reflect. He will have nothing else to do. The fruit of his reflection, of his

117

leisure, will be order. Haunting all the ages has been Aristotle's conviction that "the wise man must not be ordered but must order, and he must not obey another, but the less wise must obey him" (*Metaphysics,* Bk. I, Ch. 2, 982a). Behind Aristotle there looms Plato and beyond there is the tortured figure of Socrates before the Assembly, the good man murdered by the going democracy. Socrates knew the alternative: "one who has weighed all this i.e., the evils of the times will hold his peace, and go his own way, like a traveller in a storm of dust and sleet who seeks shelter under a wall. And as he sees lawlessness spreading on all sides, he is content if he can keep himself clear of the iniquity" (*Republic,* 496). This very ancient identificatton of Wisdom with an Order proceeding from a Philosopher who does not hold the Sword of Power and who is therefore frustrated in the drama of life was spelled out mordantly by the Platonic Socrates: "I said: until philosophers be kings, or the kings and princes of this world have the spirit and power of philosophy, and political greatness and wisdom meet in one, and these commoner natures who pursue either to the exclusion of the other are compelled to stand aside, cities will never have rest from their evils,—no, nor the human race, as I believe—and then only will this our city have a possibility of life and behold the light of day" (*Republic,* V, 474). Plato's bitter dream foresaw, even before he put the above words in Socrates' mouth, that he would be swamped by a "wave . . . of laughter and dishonor." Plato was a prophet for some twenty-five centuries. Nobody can ask more of any seer. Those centuries manifested the same sharp distinction, a sundering sword, between contemplative wisdom and political action. Contemplation has not fit men for action and total engagement in action has withered the springs of contemplation. Both orders yielded to those "commoner natures who pursue either to the exclusion of the other."

But times change after twenty-five centuries. Work is going out of fashion because it is being done so splendidly by man's technological genius. The Sword may pass to the Philosopher. Nobody else can wield it because a new technology, still but imperfectly understood even as it rapidly moves towards fullness, integrity, and being, has preempted the role of the Practical Man. Twenty years ago America was filled with stories poking good fun at the Impractical Man. Nobody repeats those stories today. They are as dead as Joe Miller's joke book. By a last and piercing irony, the joke of all jokes played upon the rationalist universe of yesterday, the practical man has been phased out: he is no longer practical because he worked so hard at it. We don't need him anymore. He is a computer and we have plenty of

118

them around. Electronics have rubbed out the distinction between action and contemplation. Power may be falling, a fruit ripe and fat into the hands of the only man who can handle it: the Philosopher possessed of a synthetic mind and sensibility capable of asking "why." Eventually he will be forced to ask with that haunted sentinel before the ruined camp of the past, Martin Heidegger, "Why is there anything at all?" Would it not seem far more reasonable that there be nothing? An older world that once avoided the question would have answered: obviously, it would be more reasonable that there be nothing at all. Quite clearly this *is* the reasonable answer. We have dozens of analytic philosophers around to tell us that "existence" is irrational and we have no reason to doubt their testimony. But things continue *to be* despite the councils of sweet reason. And today all these things are being absorbed totally within the liberty of man as he wraps the universe within the cloak of his being. They wait upon his "yes" or "no." We are entering into an Age of Metaphysics, or we stand before a barbarism undreamt of even by those distant faces of evil that forever peer down upon men and portend a final tragedy to his adventure in history.

119

Bibliographical Note

120   **GIVEN** that our study is a synthesis or an essay towards a synthesis in the philosophy of culture, this bibliography is suggestive rather than definitive, suggestive in the sense that these books have influenced us significantly. Central to our entire conception has been the Thomistic analogy of proper proportionality, and we refer the interested reader to the classical statement: Thomas De Vio Cardinalis Caietanus, *De Nominum Analogia,* ed. P.N. Zammit, O.P., Romae apud Institutum Angelicum, 1934; for a latter-day statement in English, compare Gerald B. Phelan, "St. Thomas and Analogy," *Selected Papers,* Pontifical Institute of Medieval Studies, Toronto, 1967. The metaphysical articulation of the structure of the real implicit to our discussion of Being and Meaning is formally developed in Frederick D. Wilhelmsen, *The Paradoxical Structure of Existence,* University of Dallas Press, 1970.

The following bibliography lists books in the editions, English or otherwise, which we have alluded to or drawn upon in other ways.

Aranguren, José Luis L. *La Communicación Humana.* Mexico City: McGraw-Hill, 1967.

Belloc, Hilaire. *Charles The First.* London: Cassell & Co., 1933.

Caruso, Igor A. *La Psicologia en La Crisis.* Madrid: Ateneo, 1959.

_____. *Cultura Contemporanea.* Madrid: Ateneo, 1959.

Disandro, Carlos A. *Las Fuentes De La Cultura.* La Plata: Hosteria Volante, 1965.

Eliot, T.S. *Four Quartets*. New York: Harcourt, Brace, 1949.

———. *Notes Towards the Definition of Culture*. New York: Harcourt, Brace, 1949.

———. *The Waste Land*. New York: Harcourt, Brace, 1949.

Fuller, R. Buckminster. *Education Automation*. Carbondale: Southern Illinois University Press, 1961.

Gilson, Etienne. *Terrors of the Year 2000*. Toronto: St. Michael's, 1949.

———. *Being and Some Philosophers*. Toronto: Pontifical Institute of Medieval Studies, 1949.

Guardini, Romano. *The End of the Modern World*. Chicago: Regnery, 1968.

———. *El Mesianismo en el Mito, La Revelacion y La Politica*. Madrid: Ediciones Rialp, SA., 1956.

Heers, J. *El Trabajo en La Edad Media*. Buenos Aires: Columbia, SA., 1967.

Huizinga, Johan. *Homo Ludens*. Boston: Beacon Press, 1950.

Juenger, Friedrich Georg. *The Failure of Technology*. Chicago: Regnery, 1949.

———. *Die Spiele Ein Schlüssel Zu Ihrer Bedeutung*. Frankfurt Am Main: Vittorio Klostermann, 1953.

Leavis, F.R. and Thompson, D. *Culture and Environment*. London: Chatto & Windus, 1948.

Marcel, Gabriel. *Les Hommes Contre L'Humain*. Paris: La Colombe, 1951.

Marcuse, Herbert. *Eros and Civilization*. New York: Vintage Books, 1962.

———. *One Dimensional Man*. Boston: Beacon Press, 1966.

Marias, Julian. *Modos de Vivir*. New York: Oxford University Press, 1964.

Maritain, Jacques. *Art and Scholasticism*. London: Sheed & Ward, 1946.

———. *Philosophy of Nature*. New York: Philosophical Library, 1951.

———. *Creative Intuition in Art and Poetry*. New York: Pantheon Books, 1953.

McLuhan, Herbert Marshall. *The Gütenberg Galaxy*. Toronto: University of Toronto Press, 1962.

———. *The Mechanical Bride*. Boston: Beacon Press, 1951.

121

BIBLIOGRAPHICAL NOTES

————. *Understanding Media.* New York: Signet, 1964.

————. *War and Peace in the Global Village.* New York: Bantam Books, 1968.

————. *The Medium is the Massage.* New York: Bantam Books, 1967.

Pieper, Josef. *Leisure, the Basis of Culture.* New York: Pantheon Books, 1961.

————. *The End of Time; A Meditation on the Philosophy of History.* New York: Pantheon Books, 1954.

Poulet, Georges. *Studies in Human Time.* Baltimore: Johns Hopkins Press, 1956.

Rahner, Hugo. *Der Spielende Mensch.* Einsiedeln: Johannes Verlag, 1958.

Ramuz, C.F. *What is Man?* New York: Pantheon Books, 1948.

Riesman, David J. *The Lonely Crowd.* New Haven: Yale University Press, 1950.

Rossi, Paolo. *Los Filosofos y Las Maquinas 1400-1700.* Barcelona: Editorial Labor, SA., 1968.

Selye, Hans. *The Stress of Life.* New York: McGraw-Hill, 1956.

Snow, C.P. *The Two Cultures and a Second Look.* New York: Signet, 1964.

Stallman, Robert W., ed. *Critiques and Essays in Criticism.* New York: Ronald Press Co., 1949.

Thibon, Gustave. *La Crisis Moderna Del Amor, Diagnosticos De Fisiologia Social.* Madrid: Editora Nacional, 1958.